Pharmaceutical Microbiology

Related titles

Sterility, Sterilisation and Sterility Assurance for Pharmaceuticals
(ISBN 9781907568381)

Woodhead Publishing Series in Biomedicine:
Number 80

Pharmaceutical Microbiology

Essentials for Quality Assurance and Quality Control

Tim Sandle

AMSTERDAM • BOSTON • CAMBRIDGE • HEIDELBERG
LONDON • NEW YORK • OXFORD • PARIS • SAN DIEGO
SAN FRANCISCO • SINGAPORE • SYDNEY • TOKYO
Woodhead Publishing is an imprint of Elsevier

WOODHEAD
PUBLISHING

Woodhead Publishing Limited is an imprint of Elsevier
80 High Street, Sawston, Cambridge, CB22 3HJ, UK
225 Wyman Street, Waltham, MA 02451, USA
Langford Lane, Kidlington, OX5 1GB, UK

Notices
Knowledge and best practice in this field are constantly changing. As new research and experience broaden our understanding, changes in research methods, professional practices, or medical treatment may become necessary.

Practitioners and researchers must always rely on their own experience and knowledge in evaluating and using any information, methods, compounds, or experiments described herein. In using such information or methods they should be mindful of their own safety and the safety of others, including parties for whom they have a professional responsibility.

To the fullest extent of the law, neither the Publisher nor the authors, contributors, or editors, assume any liability for any injury and/or damage to persons or property as a matter of products liability, negligence or otherwise, or from any use or operation of any methods, products, instructions, or ideas contained in the material herein.

ISBN: 978-0-08-100022-9 (print)
ISBN: 978-0-08-100044-1 (online)

British Library Cataloguing in Publication Data
A catalogue record for this book is available from the British Library

Library of Congress Cataloging-in-Publication Data
A catalog record for this book is available from the Library of Congress

For Information on all Woodhead Publishing publications
visit our website at http://store.elsevier.com/

Working together
to grow libraries in
developing countries

www.elsevier.com • www.bookaid.org

Dedication

"This book is dedicated to my wife Jenny, for her invaluable encouragement support."

Contents

Introduction

This book is concerned with pharmaceutical microbiology. Pharmaceutical microbiology is a relatively new discipline, or at least, one defined as a distinct subject matter. A trawl though textbooks pre-dating the mid-1980s reveals very few references to this sub-division of microbiology. Pharmaceutical microbiology can be simplified as being concerned with keeping microorganisms in control within the environment and hence from contaminating the product and harnessing microorganisms to make efficacious pharmaceutical medications [1]. Here pharmacists and microbiologists work synergistically to ensure that drug therapies target opportunistic microorganisms without harming its human host.

In considering the two facets of pharmaceutical microbiology, terms of product protection, pharmaceutical microbiology is concerned with minimizing the numbers of microorganisms within a process environment and excluding microorganisms and microbial by-products (such as bacterial exotoxin) from water and other starting materials. With sterile products, there is an additional requirement: ensuring that the finished pharmaceutical product is sterile. With the utilisation of microorganisms, pharmaceutical microbiology is concerned with the study of anti-infective agents (such as new antibiotics); the use of microorganisms to detect mutagenic and carcinogenic activity in prospective drugs; as well as the biotechnological application of microorganisms in the manufacture of pharmaceutical products, such as insulin.

Pharmaceutical microbiologists operate within both quality assurance and quality control. They are engaged in laboratory testing and environmental monitoring to support these activities; and acting as subject matter experts on topics as diverse as sanitization, autoclave operation, depyrogenation, process validation, selecting cleanroom gowns, specifying raw materials, and so on.

What this book does is to take pharmaceutical microbiology and place it at the heart of the pharmaceutical organization or healthcare institution. Here the book addresses a number of themes, such as drug development, the use of microorganisms in manufacturing, contamination control strategies, laboratory test methods and risk assessment.

There are a few volumes of pharmaceutical microbiology-related books available. These texts tend to focus heavily on the role of microorganisms in biotechnology or the removal or elimination of microorganisms through practices of disinfection and sterilization. Some other texts discuss laboratory test methods in isolation. What this book does is embrace these two themes, expand upon them and introduce a range of other topics that the pharmaceutical microbiologist needs to contend with. Here the author considers that the pharmaceutical microbiologist should only be spending a small amount of time in the laboratory. The establishment and qualification of test methods is important; however, the microbiologist is best served walking the plant, discussing new product development with research and development (R&D), being involved in cleanroom design and air-handling system upgrades, and, perhaps, foremost, undertaking risk assessments. The types of risk assessment that the microbiologist is required to be involved with are either an assessment of the significance of an above action level result where corrective or preventative actions (CAPAs) are employed or, more commonly, an assessment of the controls and measures in place to ensure that the above action result does not occur in the first place. In essence, being proactive rather than reactive.

This book is aimed at pharmaceutical microbiologists, with the idea of offering a different perspective on longstanding areas of testing (such a bioburden) as well as the new (rapid microbiological methods and risk assessment); at students of microbiology, pharmacy and healthcare subjects; and at personnel working in the field of pharmaceuticals, medical devices, biotechnology and healthcare, who wish to gain an understanding of the discipline or a specific area, such as disinfection or sterilization.

In terms of the structure of this book, Chapter 1 is an introduction to pharmaceutical microbiology, including a basic undertaking of microbial growth [2]. The chapter considers some of the key tests undertaken in pharmaceutical microbiology laboratories and outlines these along the path of pharmaceutical manufacturing (from starting materials to finished products). The chapter also addresses testing in support of products, utilities and the assessment of cleanrooms. Some basic matters in terms of counting plates, sampling, speciation, and risk assessment (including the necessity of a contamination control strategy) are addressed. The chapter provides a stand-alone introduction to the subject matter.

Building on the first chapter, Chapter 2 presents an outline of the pharmaceutical industry with a focus on the role of the microbiologist within it. For those readers unfamiliar with the intimate working of the pharmaceutical industry, the chapter outlines the different aspects and the process of drug development. Pharmaceutical development and manufacture are regulated through good manufacturing practice (GMP). Chapter 3 looks at GMP and compliance and, at the same time, compares and contrasts the approach of international regulators such as the US Food and Drug Administration (FDA) and the European Medicines Agency (EMA). Included in the chapter is the importance of key pharmacopeia, which provide the basis of many laboratory tests. To be effective, the pharmaceutical microbiologist needs to understand the compliance arena.

Chapter 4 looks at an oft overlook aspect, and this is with the design, set-up and management of the microbiology laboratory. As well as good design principles, the chapter considers laboratory safety and associated microbiological risks and the qualification of equipment. The chapter, with a nod towards total quality management, introduces the topic of "lean labs" and the issue of harnessing laboratory resources efficiently.

Microbiological culture media is the foundation to many aspects of microbiology [3]. Even with the dawn of rapid microbiological methods, growing microorganisms is essential (and many rapid microbiological methods are, in fact, growth based). Chapter 5 discusses the different types of culture media, variations of the testing regimes designed to release culture media into the laboratory, and the important steps that need to be followed when manufacturing culture media.

Chapter 6 continues with the microbiology laboratory and reviews the principle methods that form part of the pharmaceutical microbiologist's armory. Considered here is good laboratory practice, health and safety, aspect technique, microscopy, and culture-dependent methods. From this, the tests considered are total viable aerobic count, tests for specified microorganisms, optical density measurements, the test for sterility, pyrogen testing, endotoxin testing, and water analysis. With each test, aspects relating to method validation are highlighted, together with best practice advice.

The following chapters discuss the more important microbiological methods in more detail. Chapter 7 is concerned with bioburden testing. The chapter presents the main test methods for conducting bioburden, including plate counting (pour plates, spread plates, and membrane filtration) and the most probable number technique. An important discussion is contained within the chapter in relation to the "colony-forming unit" (what it is and what it is not) and the complexities of counting microorganisms in terms of under- and over-estimation. For readers interested in sterile products, a section is included on pre-final filtration bioburden determination. Chapter 8 is about objectionable and specified microorganisms. Although the specified microorganisms listed in the major pharmacopoeia cannot be overlooked, the emphasis in the chapter is upon risk and the importance of the microbiologist understanding the products and processes so that those microorganisms that could damage the product or cause harm to the patient are characterized and tracked. This additionally requires knowledge of the interaction between microorganisms and people, which is addressed through an introduction to the metagenomics of the Human Microbiome Project.

Understanding the types of microorganisms, as use as controls in testing, to confirm organisms used in product development or to identify contamination is a core part of pharmaceutical microbiology [4]. With contamination events and out-of-limits events, failure to characterize invariably leads to a failure to conduct an investigation soundly. Chapter 9 is about microbial identification and the chapter presents an array of phenotypic and genotypic methods. Given that choice between competing methods can often be bewildering, the chapter offers a matrix that can be adopted for selection and qualification.

Water is an important concern for microbiologists and the use of water is an inescapable feature of pharmaceutical production. Water, when produced and controlled properly, plays an effective role. However, poor design and control can lead to severe microbiological issues, not least because water acts as both a growth source and a vector for microbes. Chapter 10 looks at supplied potable water and the purification process that generates purified water and water for injections. As well as good design principles, the chapter also discuss the types of microorganisms found and requirements of testing for viable counts, bacterial endotoxin and total organic carbon.

Closely related to water is bacterial endotoxin, which is the major source of pyrogens within the pharmaceutical organization. Chapter 11 looks at the nature and sources of endotoxin (and Gram-negative bacteria) and the nature and risks of pyrogenicity. In addition, there is an in-depth review of the primary test for endotoxin: *Limulus* amebocyte lysate (LAL) method.

Chapter 12 places microbiology at the core of many production processes and examines sterility and sterility assurance. The microbiologist needs a core understanding of sterility assurance principles, whether this is for sterile products or medical devices, or for the consumables used within cleanrooms and laboratories [5]. The chapter examines the theory of sterilization and some of the different methods that can be deployed to achieve it. The theme of sterility is extended to Chapter 13, which considers biological indicators (preparations of microbial spores) used to qualify sterilization (such as moist heat, dry heat, ethylene oxide, and hydrogen peroxide). The chapter discusses concepts such as D-value, population and identification, with a focus on biological indicator resistance.

Chapter 14 examines antimicrobials. An antimicrobial is any substance of natural, semi-synthetic or synthetic origin that kills or inhibits the growth of microorganisms but causes little or no damage to the host. In undertaking this, the chapter considers two important aspects of pharmaceutical product efficacy. The first is the assessment of antibiotics to determine if the antibiotic is effective against the target microorganism. The second is the assessment of pharmaceutical products that contain preservatives, where testing is undertaken in order to assess whether the preservative can suppress the growth of any contaminating microorganisms.

Cleaning and disinfection are of great importance to process environments. Without efficient cleaning (removal of soil) then disinfectants will not work effectively, and without a robust disinfection programme, then microorganisms will be present, either numerically or in terms of problematic species, at levels that present an environmental challenge to the product [6]. Chapter 15 presents an overview of detergents and disinfectants, and discusses issues of qualification and best practice advice for the use of such agents within a GMP environment.

Pharmaceutical processing needs to take place within a controlled environment and, in most cases, there is the requirement for a classified cleanroom. Chapter 16 reviews the design requirements for cleanrooms (air handling systems and the role of air in removing contamination). The chapter also discusses clean air devices, such as isolators and safety cabinets. The chapter devotes space to environmental control and environmental monitoring, and describes the fundamentals for designing an environmental monitoring programme as part of a bio-contamination control strategy.

Many of the laboratory methods used in pharmaceutical microbiology have remained unchanged for decades (some, such as classic counting methods being largely unchanged since the nineteenth century). The twenty-first century has seen the advent of rapid microbiological methods. These methods are designed to provide a faster time-to-result and/or a more accurate assessment of microbial numbers. Chapter 17 assesses the different types and technologies of rapid methods available, and discusses how the microbiologist can be best placed to select from different (and at times competing) technologies.

The main contribution that the pharmaceutical microbiologist can make to ensuring the safety of medicines is through risk assessment. Chapter 18 presents an introduction to risk assessment and its alignment with current GMP. The chapter then considers some risk assessment tools and techniques before moving on to outline some of the main sources of contamination within the facility. The chapter then presents some examples of risk assessment tools based on surface and air contamination. These examples show that risks can be quantified, compared and contrasted.

Chapter 19 describes another important area where pharmaceutical microbiologists should be at the forefront: validation of pharmaceutical equipment. Most items of equipment are designed to be self-sanitizing or self-sterilizing, or, alternatively, this has to be carried out manually. The pharmaceutical microbiologist must be closely involved in this process. The chapter describes the essential of validation, running through installation, operation and performance qualification. The importance of contamination control is illustrated through the example of cleaning validation. In addition to validation, the microbiologist should understand the batch manufacturing process and its constituent steps. The chapter describes this process fully and pinpoints where microbiological input is required.

Microbiological testing is of little value if the data are not analyzed. Chapter 20 is about the analysis and interpretation of microbiological data. This task is not as straightforward as with other biological data because microbial data rarely conform to standardized normal distribution. The marketed skewness of microbial data either requires the data to be transformed or for alternatives methods to be used. The handling of data is discussed with a view to producing control charts. A second area that the microbiologist needs to undertake frequently is the setting of alerts and action levels. The chapter presents ways to set such limits.

An efficient microbiology laboratory must conform to accepted standards, whether this be to GMP or International Organization for Standards (ISO) or to a national accredited body. The best means to assess these standards is through regular audit. Chapter 21 is about auditing the microbiology laboratory. The chapter will be of interest to readers who need to undertake auditing and/or work in laboratories that are subject to audit. The chapter discusses the audit process and the main areas that the auditor should focus on.

The final chapter of the book, Chapter 22, acts as a conclusion. In completing this task, the chapter thematically draws together the key points made throughout the text. The title of the chapter is "Microbial Challenges to the Pharmaceutical Industry." The chapter looks at microbial risks to pharmaceuticals, including points where contamination can arise and how risk assessment is the most effective tool to avoid such contaminating events. The chapter also considers the microbial risks to process environments. In running through these sources, the chapter ties together the main issues of contamination concern.

In outlining these chapters, it can be seen that the book seeks to place pharmaceutical microbiology at the forefront of pharmaceuticals and healthcare. Medicines may be efficacious in terms of their therapeutic value but this is of little value if the medicine is contaminated, for the therapeutic value may be lost or the medicine may be dangerous to the patient. The book also seeks to unshackle the microbiologist from the bench, equip him or her with risk assessment tools and process understanding, and position the microbiologist at the heart of the drug development and manufacturing process.

References

[1] Sandle T, Saghee MR. The essentials of pharmaceutical microbiology. In: Saghee MR, Sandle T, Tidswell EC, editors. Microbiology and sterility assurance in pharmaceuticals and medical devices. New Delhi: Business Horizons; 2011. p. 1–30.

[2] Srivastava S, Srivastava PS. Understanding bacteria. Dordrecht: Kluwer Academic Press; 2003, p. 28.

[3] Sandle T. The media kitchen: preparation and testing of microbiological culture media. In: Sutton S, editor. Laboratory design: establishing the facility and management structure. Bethesda, MD: Parenteral Drug Association; 2010. p. 269–93, ISBN: 1-933722-46-0.

[4] Jimenez L. Microorganisms in the environment and their relevance to pharmaceutical processes. In: Jimmenez L, editor. Microbiological contamination control in the pharmaceutical industry. New York: Marcel-Dekker Inc.; 2004. p. 3–7.

[5] Sandle T. Practical approaches to sterility testing. J Validation Technol 2004;10(2):131–41.

[6] Sandle T. Cleaning and disinfection. In: Sandle T, editor. The CDC handbook: a guide to cleaning and disinfecting cleanrooms. Surrey, UK: Grosvenor House Publishing; 2012. p. 1–31.

Introduction to pharmaceutical microbiology

1

1.1 Introduction

Microbiology is a biological science involved with the study of microscopic organisms. Microbiology is made up of several sub-disciplines, including: bacteriology (the study of bacteria), mycology (the study of fungi), phycology (the study of algae), parasitology (the study of parasites), and virology (the study of viruses, and how they function inside cells) [1]. These broad areas encompass a number of specific fields. These fields include: immunology (the study of the immune system and how it works to protect us from harmful organisms and harmful substances produced by them); pathogenic microbiology (the study of disease-causing microorganisms and the disease process (epidemiology and etiology)); microbial genetics (which is linked to molecular biology); food microbiology (studying the effects of food spoilage); and so on [2].

The microbiological discipline of relevance here is pharmaceutical microbiology, an applied branch of microbiology (once considered as an off-shoot of industrial microbiology but now a distinct field). Pharmaceutical microbiology is concerned with the study of microorganisms associated with the manufacture of pharmaceuticals. This is with either using microorganisms to help to produce pharmaceuticals or with controlling the numbers in a process environment. This latter concern is about ensuring that the finished product is either sterile or free from those specific strains that are regarded as objectionable. This extends through the manufacturing process, encompassing starting materials, and water. Pharmaceutical microbiologists are additionally interested in toxins (microbial by-products like endotoxins and pyrogens), particularly with ensuring that these and other "vestiges" of microorganisms (which may elicit adverse patient responses) are absent from products.

Microbiological contamination becomes a problem when it results in unwanted effects occurring in pharmaceutical preparations. In drawing from risk assessment terminology, pharmaceutical microbiology centers on understanding the likelihood of product contamination arising; understanding the severity of such contamination; considering ways to minimize contamination; and, where contamination cannot be satisfactorily mitigated, using established and developing new methods to detect contamination.

To understand the severity, it is necessary to understand the type of product, its intended use, and the nature and numbers of contaminants. Microbial contamination of sterile injectable products (parenterals) presents the greatest risk, for this may lead to death of the patient, whereas with other products, aromas, off-flavors, or discolorations, caused by microorganisms, may have fewer adverse effects. Therefore, with respect to sterile products, the main concern is with any potential microbial contamination. With

Pharmaceutical Microbiology. http://dx.doi.org/10.1016/B978-0-08-100022-9.00001-3

nonsterile products (such as inhalations, tablets, oral liquids, creams, and ointments), a level of microbial contamination may be tolerated, with a greater concern centered on the presence or absence of "objectionable microorganisms." The types of microorganisms of concern depend upon the type of product and route of administration [3].

The foundation of pharmaceutical microbiology is use of culture media: these are nutrients and growth factors designed to cultivate and grow bacteria and fungi. This foundational basis is being partially eroded through the increased use of rapid and alternative microbiological methods, although even here a substantial number of these methods remain growth based. The over-riding technique that the microbiologist must apply is an "aseptic technique" (the prevention of contamination during the manipulation of samples and cultures). The growth of microorganisms and the maintenance of aseptic technique require trained and skilled microbiologists [4].

This chapter, as a way of a short introduction to the various texts in this book, outlines the essentials of pharmaceutical microbiology, many of which are discussed in detail throughout this book.

1.2   Overview of pharmaceutical microbiology

Pharmaceutical microbiology is the application of microbiology to pharmaceutical and healthcare environments. The scope of pharmaceutical microbiology is wide ranging. However, its over-riding function is the safe manufacture of pharmaceutical and healthcare preparations and medical devices. This involves risk assessment (both proactive and reactive), together with testing materials and monitoring environments and utilities.

Some of the essential tests are described in the three main international pharmacopoeias: United States, European, and Japanese. These include:

- sterility test;
- bacterial endotoxin test;
- microbial enumeration methods, such as the microbial limits test;
- tests for specified microorganisms;
- antimicrobial susceptibility testing;
- methods and limits for testing pharmaceutical grade water;
- disinfectant efficacy;
- pyrogen and abnormal toxicity tests;
- environmental monitoring;
- biological indicators.

Many of the tests associated with the above list are the subject of chapters in this book. Beyond these, pharmaceutical microbiology continues to evolve, and the book considers contemporary developments within the field.

It would be a mistake to think of pharmaceutical microbiology as confined to a range of laboratory tests. The concept of "testing to compliance" is outdated. To address contamination risks, pharmaceutical microbiology places an emphasis upon contamination control. For this reason, pharmaceutical microbiologists are involved in a number of aspects of the production process, utility supply, and cleanroom environments. This generally involves testing and the assessment of data; conducting risk assessments, either proactively or in response to a problem; and helping to design systems as part of a contamination control strategy. These twin areas of testing and control are intermixed throughout this book.

1.3 Microbiological test methods

With testing, microbiological test methods can typically be divided between:

- Qualitative methods—where the object is to ask "Is there something there?" (such methods are concerned with the presence or absence of all microorganisms or specific species);
- Quantitative methods—this is concerned with the question "How many are there?," and this centers on enumeration methods;
- Identification methods—here the focus is the question "What are they?," and the topic is the characterization and identification of microorganisms.

This section examines the work of the microbiologist in relation to the key test areas that are required to assess the pharmaceutical processing environment (Figure 1.1).

1.3.1 Product-related testing regimes

The microbiologist should be involved in establishing the sampling and testing regime in order to assess the microbiological quality of the pharmaceutical manufacturing process. This involves selecting, sampling, and testing:

- starting materials;
- in-process samples or intermediate product;
- examination of final product formulations;
- testing of the final product.

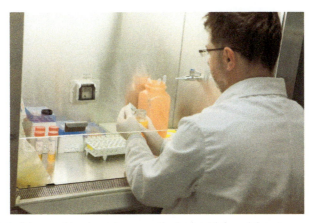

Figure 1.1 Technician working in a pharmaceutical microbiology laboratory.
Photograph: Courtesy of pharmig.

Where tests are described in pharmacopeia (of which the major global texts are the United States, European, and Japanese), then since 2000 there has been a concerted global effort to harmonize compendia [5].

1.3.2 Starting materials

The majority of starting materials (raw materials) will have pharmacopoeial monographs that will indicate the type of microbiological testing required. Depending upon the nature of the material, the intended process step, and final product requirements, additional testing may need to be considered. With a nonsterile product, for example, a presence–absence test for an objectionable microorganism not description in a pharmacopoeial monograph may be required (as discussed in Chapter 8).

Some types of starting materials may not have a supporting pharmacopoeial monograph. In such circumstances, the microbiologist will need to decide the appropriate testing regime.

The testing required on starting materials is normally divided into the following microbial limits tests:

- microbial enumeration (which is divided into tests for the total microbial count and the total yeast and mould count);
- presence–absence of specific indicator microorganisms (which may be considered objectionable to the product or process).

These two sets of tests are outlined in the United States, European and Japanese pharmacopeia (see: USP <61>/Ph. Eur. 2.6.12 "Microbiological Examination of Non-Sterile Products: Microbial Enumeration Tests" and USP <62>/Ph. Eur. 2.6.13 "Microbiological Examination of Non-Sterile Products: Tests for Specified Microorganisms").

With specified microorganisms, these are designed as indicators of potential sources of contamination. Test organisms include:

- bile-tolerant Gram-negative bacteria;
- *Salmonella*;
- *Escherichia coli*;
- *Pseudomonas aeruginosa*;
- *Staphylococcus aureus*;
- *Candida albicans*;
- Clostridia.

Not every test organism is applicable to each material. Often a pharmacopoeial monograph for a material will specify which organisms need to be examined for. Additionally, some materials required examination of endotoxin:
- Bacterial endotoxin testing.

The primary test method is the *Limulus* amebocyte lysate (LAL) test. Endotoxin testing is outlined in USP <85>/Ph. Eur. 2.6.14.

All samples should be qualified to demonstrate the ability to recover known challenges of recommended compendial organisms using the methods required (note: with the compendial methods, these methods are not required to be "validated"; however, the suitability of a given method for a particular sample must be demonstrated). Where required, representative microorganisms from the manufacturing environment can be included as a part of the assessment. With the qualification, consideration should also be given to the need for neutralization agents in the media to remove any inherent antimicrobial properties in the starting materials.

With the microbiological testing of starting materials, there may be a case for reduced testing (often in the form of skip-lot testing). This is often assessed on the basis of past test results, and the information provided by the microbiologist will be of importance in making such a case.

1.3.3 In-process samples/intermediate product

As a product is being manufactured, it is typical for in-process samples, at representative stages or points in the process where there is considered to be a risk, to be taken and submitted to the microbiology laboratory for bioburden testing. Bioburden, as Chapter 7 expands upon, is a general descriptive term for the number of bacteria living on a surface or in a material. Bioburden is normally assessed using the total viable count (TVC) method, where a portion of the sample is added to agar or agar in a receptacle is used to assess the surface of the sample. TVC provides a quantitative idea about the presence and numbers of microorganisms in or on a sample. The result obtained represents the number of colony-forming units (CFUs) per gram (or per milliliter) rather than the actual number of microorganisms present (Chapter 7 discusses this subtle, yet important, difference). Rapid microbiological methods can be used as alternatives to the pharmacopeia methods, with justification and regulatory approval [6].

For some specific process steps, such as water rinses, endotoxin testing should also be considered. In addition to samples of intermediate product, other samples can provide important information as to the risk of contamination. Such samples include granulation solutions, coating solutions, suspensions for spray drying, buffers, water rinses, and so on.

A risk assessment should be conducted for all samples submitted (to ascertain what a bioburden above a pre-defined action level means, should it occur), and this should be appropriately documented. With alert and action levels, typically there are no published limits for intermediates or in-process samples with the exception of the final bulk prior to final sterilization in aseptic manufacturing.

A further consideration is the validation of sample hold times for the microbiologist will need to establish the appropriate time between the time of sampling and testing. Typically maximum hold times are 24 h, put in place to avoid over-growth of microorganisms. The time between a sample being taken and tested should also be evaluated (during this period, samples are normally held at 2–8 °C).

1.3.4 Final product formulations

For aseptic manufacturing, it is important that the stage prior to final sterilizing filtration—(the sterile bulk) is tested. (The extent of the testing will depend upon whether the bulk is filled within a short space of time of the same site or transported elsewhere.)

As a minimum, a sample should be taken for bioburden testing. The recommended limit is 10 CFU/100 mL (for Europe this is as defined by the Committee for Proprietary Medicinal Products (as set out in CPMP QWP/486/95)) [7]. Depending upon the risk assessment, an endotoxin test is also normally conducted with a limit applicable to the final product.

Sterilizing grade filters require microbiological validation. This normally involves challenging the product with a diminutive microorganism (such as *Brevundimonas diminuta*) of a high population (typically 10^7 organisms per cm^2 of filter surface) and determining if the filter retains the microorganism or whether any of the microorganisms pass through into the filtrate [8]. Microbiologists should play a role in coordinating such studies [9].

For terminally sterilized products, it is typical for the presterilization bioburden to be assessed. This is to determine whether the microbial numbers and species types exceed the population and resistance of the microorganism used to establish the sterilization cycle. Sometimes this assessment is used as a part of parametric release in lieu of an endproduct sterility test (Figure 1.2; [10]).

1.3.5 Finished product

The finished product, for nonsterile and sterile products, must always be subject to microbiological testing. For sterile products, the testing requirements and specifications are typically defined in the pharmacopoeia. This includes the sterility test (where the product must be free from viable microorganisms) and normally an endotoxin or pyrogen test (here the limit will depend on the dose, application and intended patient type and limit calculation following the guidance in the pharmacopoeias). Some organizations will also conduct an abnormal toxicity test, using an animal model.

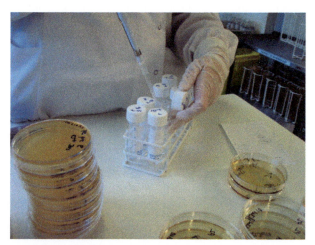

Figure 1.2 Analyst performing bioburden testing.
Photograph: Tim Sandle.

For nonsterile products, the product will typically have a specification based on the maximum permitted microbial count, and there will be a requirement for the product to be free from certain objectionable microorganisms (this is discussed later in Chapter 8). At an early stage of the product development, the microbiologist will need to be involved in establishing the specifications. This will normally be based on the preliminary testing and characterization of the product. At this stage, any requirements for neutralization of media to combat inherent antimicrobial activity of the product should also be established.

1.3.6 Testing of utilities

Pharmaceutical microbiologists are involved with the testing of various utilities that support production processes. This includes compressed gases (which are examined for microbial count and particulates), steam (which is typically examined, as condensate, for bacterial endotoxin), and, most importantly, pharmaceutical grade water. Pharmaceutical grade water includes water-for-injection (WFI) and purified water. In addition, feed water (mains water is examined). Water samples will provide an indication of the microorganisms that are circulating in the water, although they can underestimate the extent of any biofilm communities that might have formed on pipework.

Water systems are examined for:

- total aerobic microbial count;
- bacterial endotoxins;
- specific microorganisms.

With these tests, samples for microbial count are typically tested using an agar especially designed for water systems: Reasoner's 2A (R2A) agar, although other media can be used depending upon the type of water (notably, the European Pharmacopeia specifies R2A although other compendia are less specific with the selection of culture media). The method of choice is the membrane filtration method although plate counts are sometimes performed where a higher bioburden is expected from lower quality types of water. Endotoxin tests are not required for all types of water. Pharmacopoeias require the endotoxin test to be conducted for WFI and for highly purified water only. Some manufacturers also elect to test certain parts for the water system for specific microorganisms such as *E. coli* (an indicator of fecal contamination). This might include the incoming mains water, depending upon what is conducted by the water provider (such as a municipal water company). Other objectionable microorganisms can also be screened for or surveyed through the microbial identification of "out of specification" (OOS) samples.

The method for testing of water and recommended action limits are provided in the applicable pharmacopoeial monographs; however, as with environmental monitoring, it is expected that internally derived alert levels are set based upon the historical performance of the system. The frequency of monitoring depends on the state of the water system (whether it is under qualification or not) and the quality of the results. The microbiologist will need to determine suitable frequencies; here samples of WFI are typically taken each working day with the outlets on a rotational system. To achieve this, samples are taken, aseptically, at representative points of use. Other water systems are sampled less often.

The trend for a given water system or outlet is, in many ways, of more importance than individual excursions above operating limits. Trend analysis of water systems should be undertaken on a regular basis.

1.3.7   Environmental monitoring

Microbiological environmental monitoring is a key part of the assessment of pharmaceutical manufacturing facilities. Environmental monitoring data indicates if cleanrooms are operating correctly (cleanrooms are the fabric within which pharmaceutical manufacturing takes place). Other information relating to contamination control is derived from an assessment of the heating, ventilation, and air conditioning (HVAC) parameters, which contribute toward making the cleanroom "clean." These parameters include temperature, humidity, pressure, room air changes, and air-flow patterns. The microbiologist must be familiar with these.

Environmental monitoring is divided into viable and nonviable monitoring. Viable monitoring is the examination of microorganisms (bacteria and fungi) located within the manufacturing environment.

The most important element for a microbiologist to establish is the overall monitoring program. This involves the consideration of a number of factors, which include:

- the monitoring methods (to assess air, surface, and personnel contamination);
- culture media (such as general purpose media or fungal specific media) and whether a disinfectant neutralizer is required;
- incubation times and temperatures;
- sampling procedures;
- sample locations within the cleanroom;
- frequency of monitoring;
- room conditions for the sampling (at rest or in operation);
- who undertakes the monitoring (production or quality control (QC)) staff;
- establishing alert and action levels;
- guidance on dealing with out of limits results;
- characterization of the microflora.

These various considerations should be incorporated into a policy and into a plan. From this, local standard operating procedures (SOPs) should be generated. Environmental monitoring programs should adapt to changes to the pharmaceutical manufacturing environment and, therefore, the program should be regularly reviewed.

There are inherent weaknesses with environmental monitoring programs. Environmental monitoring data only provides an indication of the background environment of a cleanroom at the time it as used. A single result may or may not be representative of the longer-term room conditions. Furthermore, the methods are very insensitive, and they cannot be "validated" in the way that an analytical method can be. In addition, the frequency of samples and locations in which the samples are taken may or may not indicate the actual level of contamination risk, and there is often no direct correlation between number of organisms found and product contamination risk [11].

1.3.8 Other microbiology laboratory tests

There is a range of other tests that pharmaceutical laboratories may be involved in. These are:

- microbial identifications;
- water activity;
- disinfectant efficacy testing;
- antimicrobial susceptibility testing;
- microbial immersion studies;
- cleaning validation studies;
- maintenance of microbiological cultures;
- advising on process and equipment design;
- conducting risk assessments;
- investigation of out of limits results;
- maintenance of laboratory equipment;
- laboratory training.

1.4 The application of pharmaceutical microbiology

In undertaking the activities of pharmaceutical microbiology, there are a number of important points that should be kept in the forefront. These points range from counting methods and aseptic sampling to knowledge of objectionable microorganisms. They are presented here because they frame the assumptions of many of the tests outlined later.

1.4.1 Counting

One of the important tasks required by pharmaceutical microbiologist is the ability to enumerate microorganisms. Counting is required in order to assess the microbial quality of water, in-process bioburden samples, of raw materials, and so on. The method used to count microorganisms depends upon the type of information required, the number of microorganisms present, and the physical nature of the sample. An important distinction is between total cell count (which counts all cells, whether alive or not) and the viable count (which counts those organisms capable of reproducing).

Total cell counts include direct microscopic examination, the measurement of the turbidity of a suspension (using a nephelometer or spectrophotometer), and the determination of the weight of a dry culture (biomass assessment), adenosine triphosphate (ATP) measurements (typically using the enzyme luciferase that produces light on the hydrolysis of ATP) fluorescent staining, or electrical impedance. Viable counting techniques include the spread plate, pour plate (through direct plating or an application like the Miles–Mistra technique), spiral plating, and membrane filtration.

1.4.2   Sampling

An important aspect of the work of pharmaceutical microbiologists is ensuring that the samples taken or submitted to the laboratory have been done so in an aseptic manner and that the containers and storage conditions of the sample have not been adversely affected.

The containers selected must normally be sterile (either disposable plastic or auto-claved glassware), and the containers should not leech out any chemicals that might have antimicrobial properties and thereby give a false-negative result. The sampling technique used, irrespective of the sample type, must ensure that the hygiene of the sample is maintained (aseptic technique). This requires the person taking the sample to have been appropriately trained. The sample must also be transported in a sound manner and stored under appropriate conditions (in-process and water samples are normally required to be placed at 2–8 °C within 1 or 2 h of sampling).

In terms of the numbers of samples taken or the amount collected, the sample should be representative. This means that the sample should be of sufficient volume (such as 200 mL of pharmaceutical grade water) or an appropriate number of samples should be taken in order to produce a representative result (e.g., determining how many samples from a give number of containers of a raw material will give a repre-sentative result. There are different statistical tools which can be used for this purpose, the most simple being the square root of the number of containers). A further issue is with the sample being representative. For example, a water sample collected for microbiological analysis should be taken in the way that a production operative would use it; a raw material sample taken from a container should be of the mixed sample, so that it is homogenous; and so on. This is because microorganisms rarely follow normal distribution, and it may sometimes be that more than one sample is required for a final result to be closer to the "true" bioburden.

1.4.3   Microorganisms detected from pharmaceutical manufacturing environments

Studying the range, types, and patterns of microorganisms found in cleanrooms can provide essential information for microbiologists and quality personnel in un-derstanding cleanroom environments and for assisting with contamination control. Such studies can prove to be very useful for microbiologists in benchmarking the types and frequency of incidence of the more common microorganisms likely to occur in cleanrooms. This is an important feature of ensuring good microbiological control [12].

Two aspects of screening microbiota are to note possible resistant strains and ob-jectionable microorganisms. Where repeated occurrences of microorganisms occur in the environment, particularly Gram-positive sporing rods and certain Gram-negative rods, then this may be an indication of an ineffective cleaning and disinfection re-gime [13]. Microorganisms pose a contamination risk to pharmaceutical processing. It is important that pharmaceutical manufacturers undertake adequate cleaning and disinfection to reduce the risk of microorganisms proliferating. For this, a variety of disinfection agents may be used depending on the types of microorganisms. Prior to the purchase of new disinfectants, internal validation studies need to be undertaken in order to confirm the effectiveness of the disinfectant using recommended type cultures and "in-house" environmental microorganisms. Microbiologists should be involved with cleaning validation, where an assessment is made of the ability of an automated or manual process to remove microorganisms as well as chemical impurities [14].

The concern with objectionable microorganisms from regulators is primarily aimed at non-sterile products (creams, ointments, tablets, and so on). Many of the species considered to be "objectionable" are described by the US Food and Drug Administration (FDA) [15]. When objectionable microorganisms occur, the advice of Sutton is to conduct a risk assessment.

Risk assessment criteria may include:

(a) Absolute numbers of organisms seen.

High numbers of microorganisms may affect product efficacy and/or physical/chemical stability. An unusually high number of organisms seen in the product may also indicate a problem during the manufacturing process, or an issue with a raw material. The high bacterial counts may also indicate that the microorganisms are thriving in the product.

(b) The type of microorganisms and the characteristics of the microorganism.

The type of microorganism provides information that will indicate its probable origins and the potential risk it may pose to the product or to the environment.

(c) Considering if the microorganism can survive.

This can be based on the following factors:
- pH;
- salt concentrations;
- sugar concentrations;
- available water;
- temperature;
- time.

Survival is generally lower under the following conditions:
- low or high pH;
- high salt concentrations;
- high sugar concentrations;
- low presence of water;
- temperatures above 45 °C (except for thermophilic spores);
- low temperatures below 10 °C (although this may inhibit growth rather than destroy cells);
- freezing.

(d) Product characteristics.

The dosage form of the product is important to consider, particularly whether the material is anhydrous or water based. This can have an effect on the ability of microorganisms to proliferate. Here, whether the product contains sufficient free water to support microbial growth is a key factor.

(e) Potential impact on patients.

Assessing the risk to patients from viable microorganisms or from toxins is the aspect of risk assessment that is the most important consideration.

With both sterile and nonsterile products, objectionable microorganisms pose a concern in relation to [16]:

(a) toxins and other microbial by-products;

(b) diminutive microorganisms that may pose a challenge to sterilizing grade filters;

(c) the microbial bioburden of incoming raw materials (when assessed against the microbial limits test);

(d) as indicator organisms that may signal a concern with a utility (such as fecal coliforms in water);

(e) microorganisms with the potential to affect integrity of container/closure system;

(f) an organism that signals a concern with personnel (such as poor personal hygiene);

(g) an unusual microorganism that signals a change in the established microflora in the environment;

(h) a possible weakness with the microbial identification system (i.e., an organism that is characterized but is considered to be so unlikely to occur within a pharmaceutical environment that a misidentification may have occurred, which is sometimes possible with phenotypic techniques). Identification methods are outlined in Chapter 9;

(i) understanding the types of microorganisms provides an indication of the origin of the contamination.

1.4.4   Contamination control strategy

The tests described, and the awareness of the areas of risk throughout the manufacturing process, should feed into a contamination control strategy. This should be a formal, high-level document to which microbiologists should make a major contribution. With such documents, the foremost consideration should be that pharmaceutical medicines are to be manufactured to be safe and efficacious. The presence of microorganisms in pharmaceutical preparations can have an adverse effect on the effectiveness of the preparation, and it may cause harm to the patient. The risk of the microorganism causing harm depends upon the type of product, the way it is administered, and the health of the patient who is receiving the medicine.

1.4.5   Advances in pharmaceutical microbiology

Recent advances in pharmaceutical microbiology relate to progress made with rapid and alternative microbiological methods; and with advances in the characterization of microorganisms, which have led to a re-interpretation of microbial taxonomy.

Rapid microbiology is a reference to microbiological testing that is evolving beyond traditional methods where microorganism detection requires days or weeks to technologies that can produce a result in a much faster time. Many rapid microbiological method technologies provide more sensitive, accurate, precise, and reproducible test results when compared with conventional, growth-based methods. Furthermore, they may be fully automated, offer increased sample throughput, operate in a continuous data-collecting mode, and provide significantly reduced time-to-result (sometimes in "real-time").

Rapid microbiological methods and alternative methods are often used as interchangeable terms, although more strictly "alternative methods" refers to techniques that differ to those described in compendia. Many rapid and alternative methods can also detect microorganisms that are present in a sample (or within an environment) but that cannot be easily cultured. This is because the microorganisms are either stressed or sub-lethally damaged or because they simply cannot grow on standard culture media. Such microorganisms are referred to as "viable but non-culturable" or "active but non-culturable" [17]. Rapid methods are discussed in Chapter 17.

Microbiological understanding of people and how they relate to the surrounding world has advanced considerably following the publication of the key findings from the first wave (2008–2013) of the Human Microbiome Project. These findings not only confirmed that the human body is an intricate system that hosts trillions of microbial cells, across the epithelial surface, and within the mouth and gastrointestinal tract; it also demonstrated that microorganisms play a complex role in human physiology and organ function, influencing digestion, immunity and development [18]. The microbial community also affects the way that different medicines work in the body, leading to possibilities of medicines tailored for different individuals based on the genetic interactions between the individual and their microbial communities.

The Human Microbiome Project has also directed the development of new genotypic and molecular methods (the most prevalent methods use comparative deoxyribonucleic acid (DNA) sequencing of the 16S ribosomal ribonucleic acid (rRNA) gene in bacteria and a region associated with the 26S rRNA gene in fungi). The use of these methods has led to advances with identification and the re-classification of the bacterial and fungal kingdoms [19]. Moreover, metagenomics (the study of genetic material recovered directly from environmental samples) has led to a greater understanding of how microorganisms interact with, and have an influence on, their hosts and the surrounding environment.

1.5 Conclusion

This chapter is an introduction to the more detailed aspects of pharmaceutical microbiology that are discussed throughout the rest of this book. The purpose of the chapter was to provide an overview of some of the specifics of pharmaceutical microbiology. Pharmaceutical microbiology covers a very large area and this chapter can only touch on some of the more common and essential elements. In doing so, the chapter has bridged the laboratory testing side of pharmaceutical microbiology with the contamination control side. These sides are sometimes inelegantly split between "QC" and "quality assessment" microbiology. This is unfortunate, for microbiology should not be artificially separated and instead each organization should have in place a site microbiologist who can keep both parts focused on contamination control.

References

[1] Hodges N. Fundamental features of microbiology. In: Denyer SP, Hodges N, Gorman SP, Gilmore B, editors. Hugo and Russell's pharmaceutical microbiology. 8th ed. London: Wiley-Blackwell; 2011. p. 15.

[2] Black JG. Microbiology: principles and applications. 3rd ed. Upper Saddle River, NJ: Prentice Hall; 1996, p. 136–40.

[3] Richter S. Product contamination control: a practical approach bioburden testing. J Validation Technol 1999;5:333–6.

[4] Tortora GJ, Funke BR, Case CL. Microbiology: an introduction. 5th ed. Redwood City, CA: The Benjamin/Cummings Publishing; 1995, p. 155–8.

[5] Sutton SVW, Knapp JE, Porter D. Activities of the USP Analytical Microbiology Expert Committee during the 2000–2005 revision cycle. PDA J Pharm Sci Technol 2011;59(3):157–76.

[6] Clontz L. Microbial limit and bioburden tests: validation appropriates and global requirements. New York: CRC Press; 2009, p. 1–10.

[7] The European Agency for the Evaluation of Medicinal Products Committee for Proprietary Medicinal Products (CPMP) note for guidance on manufacturing of the finished dosage form, CPMP/QWP/486/95. London, UK: Human Medicines Unit.

[8] Akers JA. Microbial considerations in the selection and validation of filter sterilisation. In: Jornitz MW, Meltzer TH, editors. Filtration and purification in the biopharmaceutical industry. New York: Informa Healthcare; 2008. p. 151–61.

[9] Jornitz MW, Akers JE, Agalloco JP, Madsen RE, Meltzer TH. Considerations in sterile filtration. Part II: the sterilising filter and its organism challenge: a critique of regulatory standards. PDA J Pharm Sci Technol 2003;57(2):88–95.

[10] Sandle T. Sterility, sterilisation and sterility assurance for pharmaceuticals. Cambridge, UK: Woodhead Publishing; 2013, p. 119–20.

[11] Noble NH. Validation of environmental monitoring methods and levels. PDA J Parenter Sci Technol 1993;47:26–8.

[12] Farquharson G. Clean rooms and associated controlled environments. Pharm Technol Eur 2002;14:43–7.

[13] Sandle T. Gram's stain: history and explanation of the fundamental technique of determinative bacteriology. IST Sci Technol 2004;54:3–4.

[14] Hall WE. Validation and verification of cleaning processes. In: Nash RA, Wachter AH, editors. Pharmaceutical process validation. 3rd ed. New York: Marcel Dekker; 2003. p. 465–506.

[15] Cundell AM. Microbial testing in support of aseptic processing. Pharm Technol 2004;28:56–64.

[16] Sutton S. Microbial limits tests: the difference between "absence of objectionable microorganisms" and "absence of specified microorganisms". PMF Newsl 2006;12(6):3–9.

[17] Cundell AM. Opportunities for rapid microbial methods. Eur Pharm Rev 2006;1:64–70.

[18] Cundell AM. Implications of the human microbiome project to pharmaceutical microbiology. In: Griffin M, Reber D, editors. Microbial identification: the keys to a successful program. Bethesda, MD: PDA/DHI; 2012. p. 399–406.

[19] Moreno-Coutino G, Arenas RF. Fungi: types, environmental impact and role in disease. In: Silva AP, Sol M, editors. Fungi: types, environmental impact and role in disease. New York: Nova Biomedical; 2012. p. 41–58.

Microbiology and pharmaceuticals

2

2.1 Introduction

Pharmaceuticals represent a relatively small sector of manufacturing industry (and in hospitals, pharmacy units are a small part of the hospital function). Nonetheless, the pharmaceutical sector is highly innovative, closely regulated, and (in the private sector) very profitable. At some time or another, most people will use the products of the pharmaceutical industry. These range from life-saving drugs for the treatment of serious acute conditions, to drugs that can transform the life of patients with chronic, debilitating diseases, to proprietary medicines that can be bought over-the-counter (OTC; nonprescription) to relieve minor ailments. To add to this, a more recent development is the trend toward personalized medicines (for which an understanding of the human microbiome is essential). Drug expenditure varies globally, with high-income nations spending considerably more than low-income nations.

Pharmaceutical companies are chemical synthesis plants (manufacturing bulk ingredients), finished product facilities that process bulk ingredients or facilities that undertake both functions. In addition to this, there are hospital pharmacy units that produce smaller quantities of medicinal products, often on a named patient basis.

The pharmaceutical industry is global and covers everything from high technology companies, which have been founded to apply the very latest technology to the medical problems of today and tomorrow, to organizations that use relatively unsophisticated production methods to produce old, but needed, drugs at low cost. An example of the latter would be a tablet facility producing aspirin (acetylsalicylic acid).

This chapter introduces the key aspects of the industry, and later goes onto to show how microbiologists play a key role in the development of medicinal products and in ensuring that the products are safe and efficacious.

2.2 The basics of the pharmaceutical sector

The pharmaceutical sector can essentially be broken down into two main areas:

(a) proprietary;
(b) generics.

The sector can then be subdivided into two major distribution channels:

(a) prescription-only medicines;
(b) OTC medicines.

Proprietary (sometimes, without irony, called "ethical") medicines, are primarily those medicines that, because of their potency, or their ability to induce adverse effects, must

Pharmaceutical Microbiology. http://dx.doi.org/10.1016/B978-0-08-100022-9.00002-5

be taken under the supervision of a doctor and, therefore, are not available to the general public other than through a medical professional's prescription [1].

Generic products are a subgroup of prescription products. They are medicines produced by a variety of manufacturers after the original inventor's patent monopoly has expired. They are usually identical to the original product, except perhaps for minor differences, for example, in tablet color or shape, but are significantly cheaper, because the manufacturers do not bear the burdensome research and development (R&D) costs. An example here is ibuprofen.

OTC medicines are those medicines that, because of their record of safety, can be sold directly to the public and can be used without medical supervision. Aspirin is a good example of a very useful medicine that can be bought without prescription.

The pharmaceutical market is mainly dominated by the three major developed markets of the United States, Japan, and Europe (EU), although there has been substantial growth in China. Manufacturing hubs include India and Eastern Europe. The research-based, prescription pharmaceutical companies dominate the pharmaceutical industry in size, turnover, and number of employees. However, the industry is highly fragmented, and no single company is dominant. Nonetheless, mergers and acquisitions have taken place to an increasing extent amongst large companies as they seek to increase their market share, coverage of the worldwide market or gain access to new technology and therapeutic areas.

2.2.1 Pharmaceuticals

A pharmaceutical drug is a drug used in healthcare, selected on the basis of the intended action of the drug through pharmacology. Pharmaceutical medications aid the diagnosis, cure, treatment, or prevention of disease. Pharmaceuticals are classified either according to their origin or in terms of their treatment. Most pharmaceuticals are low molecular mass organic chemicals (produced by chemical synthesis); although some, such as aspirin, are isolated from biological sources. Others are of natural origin.

Variations include manufacture from biological sources, such as blood or hormones. These generally fall within the definition of "biologics" or biopharmaceuticals.

Biologics refers to pharmaceutical products derived from biotechnological processes, where cells, tissues, or biological molecules are used to manufacture the product. "Biotech" draws on a number of interdisciplinary processes including chemical engineering, bioprocess engineering, bioinformatics, and biorobotics. One key example of biotechnology is with the designing of microorganisms to produce antibiotics.

Biopharmaceuticals (alternatively biologic medical product) is a term ordinarily reserved for genetic engineering or hybridization technology. Examples of biopharmaceuticals include vaccines, blood or blood components, allergenics, somatic cells, gene therapies, tissues, recombinant therapeutic protein, and living cells. Many of these products are heat-liable sterile products; for this reason they are produced aseptically [2].

Pharmaceuticals are produced by pharmaceutical companies of varying sizes, compounding units, and within the hospital sector.

2.2.2 Product evolution

Research-based pharmaceutical companies produce new medicines that will be first marketed as prescription products. When first introduced into the market, a new product will be carefully monitored to determine if the assessments of its safety, made prior to authorizing its release on to the market, are borne out in wider usage. Such assessments are usually made at the national level (or intra-national level in the case of the European Union).

In addition, in order to recover the sizable development costs that are incurred in pharmaceutical R&D, research-based companies need the period of monopoly provided by the patent system. After some period on the market, the originator's patent will expire. This opens the way to other manufacturers to produce copies of successful products. Unlike the original medicine, which is usually sold on its brand name, these copies are usually sold, at a far lower cost, by their chemical or generic name.

It is also possible that, through widespread use, a drug may come to be recognized to be safe within broad limits. Although "safe" is never an absolute state as far as medicines are concerned, it is possible to achieve levels of safety that warrant the sale of approved formulations directly to the general public OTC without having first to obtain a prescription from a doctor. OTC products are strongly branded, and this gives them a proprietary life that can far exceed the patent life. These products also require authorization by the country regulatory bodies (e.g., in the United Kingdom, this would be the Medicines and Healthcare products Regulatory Agency; in the United States, this would be the Food and Drug Administration) and a number of further criteria, in addition to product safety, must be assessed.

2.2.3 Importance of R&D

Most industries conduct R&D at some level to improve existing products or services or to introduce new ones. As judged by R&D expenditure as a proportion of sales, the pharmaceutical industry is one of the high spending industrial sectors (exceeded, perhaps, only by computers and other electronic devices). R&D spending in major pharmaceutical companies typically lies in the range 15–20% of sales [3].

R&D is arguably the key driver in propelling forward the growth of the pharmaceutical industry, through the release of new products together with innovations to existing products. R&D is an ambiguous term. The meaning varies with the context in which it is used; for example, R&D for the insurance industry is entirely different to R&D for the aerospace industry. In the context of the pharmaceutical industry, *research* is the process that investigates disease mechanisms and substances that modulate the disease and which, if successful, will identify a substance that is considered to have the potential to become a drug. *Development* is what turns that substance into a marketed medicine that doctors can prescribe and that will be of benefit to patients.

2.2.4 Development

Drug development is the process of turning "an interesting compound" into a marketed product. It is partly concerned with proving the safety and efficacy of the chosen compound to get marketing approval from the regulatory authorities. It also includes other activities that are essential for a marketed product, such as the manufacturing process, the shape and color of the tablet (if that is the chosen dosage form), and how the product will be packaged. Health and safety data, in addition to the safety data for the compounds as a drug, will be required to guide staff who will be working with the chosen compound. The "commercial profile" of the compound will become apparent during its development; it will need to have attractive indications and claims that will allow it to compete with other treatments for the target indications, including nondrug treatments if the new product is to be commercially successful.

Drug development is expensive. Each drug development project will be unique and its cost will depend on many factors, the main ones of which are:

- The cost of synthesis of the chemical, biological, or recombinant molecular entity that has been selected for development, especially when the manufacturing processes must comply with good manufacturing processes;
- The extent and nature of the toxicology program, for example, whether lifetime animal studies are needed to determine the drug's carcinogenic potential;
- The extent of the clinical program. Some drugs are approved with clinical data from a few hundred patients, but more usually several thousand will be involved in a clinical program;
- The complexity of the clinical endpoint in the studies needed for approval (and marketing). Trials with a new antibiotic in which the patient's return to health after a short course of treatment for an uncomplicated infection and the proven eradication of infecting organisms, for example, in a urine sample, is a simple and clear-cut clinical endpoint. This is to be contrasted with trials in patients who are very sick and have a number of concomitant medical conditions that require treatment at the same time as the indication under study. For example, patients with septic shock, or trials in patients that need to show that the drug being studied will increase a patient's life expectancy, as is sometimes the case with drugs intended to treat cancers or heart disease.

2.2.5   Cost

Prices of new pharmaceuticals are often relatively high, especially when the ingredient and manufacturing costs contribute only a small fraction of the retail price (sometimes less than 10%). The industry justifies its new drug prices on the grounds that they are needed to give an adequate return on the massive investment made in R&D over the (relatively) short period of patent-protected market exclusivity. Prices of new drugs are directly controlled by the state in a number of countries, which has the effect of creating a new regulatory step in the approval process to determine pricing.

2.2.6   Generic drug products

The pace of pharmaceutical innovation is relatively slow, and the long development periods mean that the period of market monopoly is sometimes only a few years. This means that generic copies of "state-of-the-art" therapies can sometimes appear within a few years after the first launch of a new medicine.

Related to generics is compounding. Pharmacy compounding is the process of preparing personalized medications for patients. Compounded medications are "made from scratch," where individual ingredients are mixed together in the exact strength and dosage form required by the patient. This method allows the compounding pharmacist to work with the patient and the prescriber to customize a medication to meet the patient's specific needs.

2.2.7   Other sectors

Like any other major industry, the pharmaceutical industry has spawned a number of satellite sectors. This includes contract manufacturers, contract research organisations (CROs), companies specializing in formulation technology and biotechnology.

Contract manufacturers are companies that specialize in some aspect of pharmaceutical manufacturing technology and that other companies may contract to undertake some aspect of manufacture on their behalf. Firms in this category can range from fine chemical manufacturers, through formulation processors to specialist-packing companies. Contract companies can use conventional technology providing, for example, tableting capacity or sterile product facilities, or they can be developing new technology, as in the companies specializing in chiral synthesis.

Contract manufacturers may be contracted on a short-term basis to fulfil a one-off sales contract or on a long-term basis, providing, for example, manufacturing capacity in territories where it is desirable to manufacture within the country concerned. They also represent an alternative for research-based firms that may require new manufacturing facilities for a new product. A contract manufacturer can provide manufacturing capacity while the new product is being established in the market allowing the inventing company to delay making a capital investment until the future of the product is more secure and predictable.

CROs provide clinical services for drug development; they conduct clinical trials. A CRO may be contracted by a major company to conduct some specific studies as a part of an overall drug development program, or smaller companies may entrust them with a clinical development program in its entirety [4].

CROs have become commonplace for two reasons. First, the bigger companies are seeking to rationalize their operations and are outsourcing many of the activities that had been carried out within a company. While for big companies, clinical development is likely to remain a core function, it is no longer necessary to staff a clinical research department to a level that would be capable of dealing with unanticipated peaks of activity. Companies are now more prepared to use CROs to deal with these peaks. The second factor has been the appearance and growth of small pharmaceutical companies, particularly biotechnology companies, which may be based on a new technology and have one or two products in development. Such companies cannot afford to set up and staff a full-time clinical function, at least initially, and the CRO is an obvious choice [5].

Formulation technology companies are another part of the pharmaceutical sector. New technology allows drugs to be formulated so that their release characteristics can be modified for their intended use, for example, the introduction of an implantable device that will release a drug at a uniform rate over a period of months. Furthermore, the life of old products can be extended by making them more convenient to administer; there are many examples of drugs that have been reformulated in "sustained release" formulations that make possible the once daily dosing of a drug that originally needed to be taken two or three times a day. Many of these developments have originated in small companies devoted to formulation technology rather than in the formulation development laboratories of the major companies.

Biotechnology has developed a wide range of technology relevant to the pharmaceutical industry. For some years, there was a widespread view that the proper place of biotechnology was in providing a better understanding of disease processes, which would enable conventional pharmaceutical research to invent better medicines. The launch of a number of very successful biotechnology products has shown that this sector of the pharmaceutical industry is of growing importance in its own right rather than just an enabling technology.

Biotechnology involves the use of microorganisms and biological processes, the most common example of which is fermentation. One application of fermentation is with the production of antibiotics, hormones, and enzymes, which are produced via the breakdown of an organic substance.

Fermenters are designed to contain an internal environment for the optima metabolization and efficient reproduction of microorganisms. They are also designed in a way to avoid contamination, beginning as aseptic spaces. With the operation of fermenters, there are a number of key operational aspects [6]:

(a) the supplied nutrients must meet the prerequisites for microbial growth;
(b) if the process is aerobic, then filtered air must be supplied;
(c) temperature must be regulated (via a thermostat);
(d) pH must be at an optimum level;
(e) mixing must be controlled; here bubble agitation is used to control nutrients and oxygen.

Such devices are either designed for batch culture (where microbial growth is halted, to allow for the removal of the required product) or continuous (where microbial growth is maintained at the exponential level and the product is continuously removed).

Thus, the biotechnology sector provides a vehicle for the rapid transfer of new technology from academia into a commercial environment. This serves not only biotechnology in the strict sense of the word but also related technologies such as computer-aided drug design, innovative approaches to formulation, and natural product screening.

2.3   Role of the microbiologist

Having outlined the basic patterns of the pharmaceutical sector, it is important to examine the oft overlooked contribution of the pharmaceutical microbiologist. Due to the diverse nature of pharmaceuticals and healthcare, the role will vary across the sector. Nevertheless, the microbiologist should play a key role within the organization. This is not least because microbial contamination in pharmaceutical products can have significant consequences. It is not only important for compliance with standards but also reduces risk to the end user, and, consequently, to the manufacturer.

Microbiologists also play an important role in drug development. Understanding the principles of microbiology and human cell mechanisms allows pharmacists to discover antimicrobial drugs that would prevent an escalating number of communicable diseases. Pharmacists and microbiologists work synergistically to ensure that drug therapies target the opportunistic microorganisms without harming its human host. Another important role in pharmaceuticals is the use of microbes for the medically important studies. For example, in the development of Bacteriorhodopsin, a protein from the plasma membrane of *Halobacterium salinarum*.

Microorganisms are used to manufacture many types of biological drug products. Microorganisms produce a variety of secondary metabolites, some of which have been identified as having therapeutic value. A prime example is with antibiotics (low-molecular-weight metabolites that inhibit the growth of microorganisms), which can be sourced from microorganisms such as *Streptomyces*, *Cephalosporium*, *Aspergillus*, or *Pencillium*, or produced artificially. For example, species of the fungus *Penicillium* produces the antibiotic penicillin (an antibiotic with a 6-amninopenicillanic acid core ring structure; branded examples include phenethicillin, propicillin, and oxacillin). Other examples include tetracyclines such as streptomycin (from species of *Streptomyces*) and cephalosporins (from the fungus *Cephaloporium acremanium*) [7]. Further aspects of pharmaceutical microbiology include the R&D of anti-infective agents; the application of microorganisms to detect mutagenic and carcinogenic activity in prospective drugs; and the use of microorganisms in the manufacture of products such as insulin and human growth hormones.

Microbiology plays a significant role in medical devices, such as fluorescent fusion, which are used for fast and precise detection of pathogens in tissue samples. It is a technology for carrying out immunofluorescence studies that may be applied to find specific cells in complex biological systems [8].

There is also a part played by microbiologists in the cosmetics industry. This is because the contamination of cosmetics can result in them being converted into products hazardous for consumers. The water and nutrients present in cosmetics make them susceptible to microbial growth, although only a few cases of human injury due to contaminated cosmetics have been reported. More often, microorganisms are the cause of organoleptic alterations, such as offensive odours, and changes in viscosity and color.

Pharmaceutical microbiologists provide an essential contribution to risk assessment. This can be with assessing new systems and processes to determining where contamination risks may occur (linking with "Quality by Design" concepts). Risk assessment also applies to reviewing contamination events for their significance. With these two facets of risk management, proactive risk assessments should take up the greatest proportion of time. It follows that the more effectively this is done, then the less likely contamination situations are to occur. Tools for performing such assessments include risk analysis tools borrowed from other industries or professions including HACCP (hazard analysis critical control points) from the food industry, FMEA (failure modes and effects analysis), and FTA (fault tree analysis) taken from engineering industries, such as, car production. These approaches share a number of things in common:

- constructing diagrams of work flows;
- pin-pointing areas of greatest risk;
- examining potential sources of contamination;
- deciding on the most appropriate sample methods;
- helping to establish alert and action levels;
- taking into account changes to the work process/seasonal activities.

In order to assess these aspects, it is important that the microbiologist builds up detailed knowledge of the production system and processes. Thus, the lead microbiologist will spend little time testing and accumulating data and more time formulating corrective and preventive actions and performing process reviews [9].

Risk assessment is also integral to drug development. Here the most important intent in risk determination is, of course, patient safety. Second intent is the risk to the product, to prevent batch rejection. Of these two, the patient's risk must always be the highest concern. Patients depend on products to be safe from contamination when they first use them and if they continue to use them.

A set of three parameters can be used to develop an understanding of the risks related to patient safety. These are:

- Route of administration—this concerns how the drug will be delivered to the patient: method, mechanism, or through which part of the human body will the drug impact the most;
- Patient health—this is the real time health of a patient receiving the drug and this is an aspect of age, culture, and pre-disposed conditions that could be environmentally imposed;
- Target dose intent—this concerns the target organ, region, or breadth of exposure to the human body for the pharmaceutical product.

Risk assessment concepts are discussed in more detail in Chapter 18.

It is not sufficient for the pharmaceutical microbiologist to have knowledge of microbiology. The microbiologist needs to have a wider knowledge of physics [10]. Without such an appreciation, the significance of results from a cleanroom, whether viable microorganisms or nonviable particles, cannot be fully understood. Physical tests, such as pressure differentials, clean-up times (recovery rate), and airflows frame the context of the microbiological result. Likewise, the microbiologist is required to have a greater understanding of chemistry, engineering, and engineering systems. For example, in assessing the results from a purified water system, some knowledge of flow rates, valve design, re-circulation, heating, and piping is required. To add to this, an understanding of the chemistry of disinfectants and antiseptics is necessary in order to select and monitor the most potent and effective compounds.

2.4 Conclusion

The purpose of this chapter was to provide an introduction to the pharmaceutical industry so that the importance of, and complexities involved with, drug development could be outlined. The cost and risks of introducing new drugs are high, and pharmaceutical microbiologists are frequently required to play a part in this development process, including the establishment of test methods and the setting of test limits.

The role of the microbiologist also extends to mapping out contamination control strategies, undertaking risk review and assessing contamination control problems. It also extends to designing and executing an array of tests. In short, the microbiologist plays a critical role in the pharmaceutical development process.

References

[1] Giannakakis IA, Ioannidis JPA. Arabian nights—1001 tales of how pharmaceutical companies cater to the material needs of doctors: case report. BMJ 2000;321:1563–4.

[2] Walsh G. Biopharmaceuticals: biochemistry and biotechnology. 2nd ed. Chichester, UK: Wiley; 2003, p. 1–2.

[3] Lexchin J, Bero LA, Djulbegovic B, Clark O. Pharmaceutical industry sponsorship and research outcome and quality: systematic review. BMJ 2003;326:326–35, http://dx.doi.org/10.1136/bmj.326.7400.1167.

[4] Ioannidis JPA, Karassa FB. The need to consider the wider agenda in systematic reviews and meta-analyses: breadth, timing and depth of evidence. BMJ 2010;341:341–46, http://dx.doi.org/10.1136/bmj.c4875.

[5] Lathyris DN, Patsopoulos NA, Salanti G, Ioannidis JPA. Industry sponsorship and selection of comparators in randomized clinical trials. Eur J Clin Invest 2010;40:172–82.

[6] Hogg S. Essential microbiology. Chichester, UK: Wiley; 2005, p. 133–6.

[7] Wainwright M. Moulds in ancient and more recent medicine. Mycologist 1989;3(1):21–3.

[8] Sandle T. The changing role of the pharmaceutical microbiologist. PharMIG News 2003;(12):2–3.

[9] Sandle T. Risk management in pharmaceutical microbiology. Pharm Manuf 2011;10(8):30–6.

[10] Sandle T. Recent developments in European regulatory requirements: issues affecting the microbiologist. Pharm Microbiol Forum Newsl 2012;18(1):5–15.

GMP and regulations

3

3.1 Introduction

The pharmaceutical industry is highly regulated by the application of the principles of good manufacturing practice (GMP). In most countries, government agencies provide guidance to pharmaceutical manufacturers that is intended to facilitate the manufacture of safe, unadulterated and efficacious drug products. The pharmaceutical industry is one of the most highly regulated, and regulation is enforced by governmental and international agencies [1].

The sterile pharmaceutical sector has a well-defined set of expectations and regulations that provide clear statements relating to microbiological controls and monitoring. In contrast, the expectations for nonsterile pharmaceuticals are poorly defined, with few specifics written in either legislation or guidance publications. The regulatory agencies, therefore, expect the industry to take a risk-based approach to microbiological control and apply appropriately justified monitoring in the manufacture of nonsterile pharmaceuticals [2]. This chapter outlines the key requirements for GMP that can be applied across the pharmaceutical industry.

With microbiological aspects of GMP, top level microbial oversight "governance" should be driven by senior level site management and not limited to the senior microbiologist or quality assurance (QA) manager. Clear direction needs to be given to the site emphasizing that microbiological contamination control is a key factor in GMP. Multifunctional involvement of representatives from manufacturing (technical and operations), engineering, QA management, and quality control (QC) microbiology should have a collective responsibility to ensure the appropriate quality systems are in place to ensure microbiological control.

3.2 Good manufacturing practice

GMP refers to the rules governing the manufacture of a safe and efficacious pharmaceutical product. There are two main global bodies that oversee GMP. These are the US Food and Drug Administration (FDA), where manufacturers are governed by the Code of Federal Regulations (CFRs), in particular 21 CFR 210–211, and the European Union (EU GMP), which is overseen by the European Medicines Agency (EMA). To complement this, the World Health Organization (WHO) has a GMP system, although this, to an extent, draws upon EU GMP [3]. There are, additionally, national GMP systems operative in most countries. Here there are invariably different nuances that add an extra dimension to the regulatory process [4].

Pharmaceutical Microbiology. http://dx.doi.org/10.1016/B978-0-08-100022-9.00003-7

The different regulatory agencies control the production of medicines through various licences, such as:

- manufacturers licence;
- manufacturers specials licence;
- marketing authorization licence;
- wholesale dealers licence;
- investigational medicinal products (IMPs) licence.

Compliance with these licences is assessed routinely through GMP inspections, which are normally conducted every 2 years (although with "risk-based" inspections this frequency can alter). Regulators aim to assess whether the pharmaceutical company is manufacturing the product in the way that is stated in licences, policies, procedures, and other official documentation.

Not all aspects of GMP are written down in regulations, such as innovations relating to the latest technologies. This part of GMP is called "current" or cGMP. It is up to each pharmaceutical organization to be familiar with the current "hot topics." Connected with GMP is good distribution practice (GDP), focused on the distribution of medicines; good clinical practice (GCP), which is concerned with clinical trials; and good laboratory practice (GLP), where the focus is animal experimentation. GLP should not be confused with GCLP (good control laboratory practice). GCLP, which covers QC, is a subset of GMP. Collectively, these different aspects of best practice are commonly abbreviated to GxP [5].

The two dominant sets of GMP guidance are those that relate to the EU (operated by the EMA and national agencies) and the US CFR, which form the basis of inspections by the FDA.

3.2.1 EU good manufacturing practice

The principles and guidelines of GMP are set out in two EU Directives: 2003/94/EC covers human-use medicines and 91/412/EC veterinary-use medicines. The texts of these Directives are included and expanded on in the EU GMP guide, which is contained in Chapter 4 of EudraLex on the European Commission. This guide covers GMP for medicinal products and also for starting materials used as active ingredients [6].

3.2.2 FDA and CFRs

The FDA's legal authority to regulate both medical devices and electronic radiation-emitting products is the Federal Food Drug & Cosmetic Act (FD&C Act). The FD&C Act contains provisions, that is, regulatory requirements, which define the FDA's level of control over these products. The CFR is a codification of the general and permanent rules that are published by the executive departments and agencies of the Federal Government. It is divided into 50 titles that represent broad areas subject to Federal regulation.

Most of the FDA's medical device and radiation-emitting product regulations are in Title 21 CFR Parts 800–1299. These final regulations codified in the CFR cover various aspects of design, clinical evaluation, manufacturing, packaging, labeling, and postmarket surveillance of medical devices.

3.2.3 Key aspects of GMP compliance

There are many important parts of GMP compliance. GMP has five main attributes:

- safety,
- identity,
- strength,
- purity,
- quality.

Of these, the most critical are [7]:

- Proper documentation and records—"if it is not recorded it never happened" according to the inspectors. It is important that all actions, events, and decisions relating to the quality of the product must be recorded at the appropriate level of detail in a controlled way;
- Control of materials. This refers to ensuring that all materials used, whether they be the raw materials, components such as bottles or stoppers, and packaging materials, are of the sufficient quality and are traceable;
- Thorough housekeeping and cleaning. GMP requires that people work in an orderly and methodical way and that work areas are neat, tidy and, there is segregation between tasks where required. This will reduce the potential for errors and mix-ups to occur;
- Responsible personnel behavior. This includes such areas as reporting incidents and errors immediately, and behaving appropriately in controlled areas (such as minimizing particles and microbial contamination in cleanrooms);
- Process control at all steps. This level of control relates to ensuring that all parameters are in control throughout the manufacturing process (e.g., time, temperature, pH) and reporting immediately if there is a noticeable drift or adverse trend;
- Maintenance of equipment. This involves ensuring all equipment used in the manufacture of product is "fit for purpose" and is cleaned, maintained, calibrated, and verified as appropriate and labeled/recorded as such. This is supported through initial and on-going validation. Any equipment not fit for purpose should ideally be removed or clearly labeled [8].

3.2.4 Ten rules of GMP

The essential elements of GMP can be conveyed to personnel in a way that is easy to understand as a useful training aid. GMP requires that initial and on-going training must be provided for all personnel whose duties take them into production areas or into controlled laboratories (including the technical, maintenance, and cleaning personnel), and for other personnel whose activities could affect the quality of the product.

Ten suitable GMP "rules," to be used for staff training or to act as a reminder, are set out below.

1. Confirm you are trained and have correct written instructions before starting any job?
2. Follow instructions exactly.
3. Report errors and bad practices immediately.
4. Ensure you have the right materials before you start a job.
5. Use the correct equipment for the job, confirm its status and cleanliness.
6. Maintain good segregation. Protect against contamination.
7. Work accurately, precisely, and methodically.
8. Maintain good standards of cleanliness and tidiness.
9. Ensure changes are pre-approved (through the change control system).
10. Do not make assumptions—check it out.

3.2.5 Risk management

An important part of GMP is risk management. On August 21, 2002, the FDA announced a significant new initiative: "Pharmaceutical Current Good Manufacturing Practices (CGMPs) for the 21st Century: A Risk-Based Approach." The objective was to enhance and modernize the regulation of pharmaceutical manufacturing and quality. The methodology was to use risk-based and science-based approaches for regulatory decision-making throughout the entire life cycle of a product. This initiative set forth a plan to enhance and modernize the FDA's regulations governing pharmaceutical manufacturing and product quality for human and veterinary drugs and human biological products [9].

The objectives were to:

• encourage the early adoption of new technological advances by the pharmaceutical industry;
• facilitate industry application of modern quality management techniques, including implementation of quality systems approaches, to all aspects of pharmaceutical production and QA;
• encourage implementation of risk-based approaches that focus both industry and FDA attention on critical areas;
• ensure that regulatory review, compliance, and inspection policies are based on state-of the-art pharmaceutical science;
• enhance the consistency and co-ordination of the FDA's drug quality regulator programs, in part, by further integrating enhanced quality systems approaches into the Agency's business processes and regulatory policies concerning review and inspection activities.

The risk approach has been adopted by other regulators and reached international agreement through the documents issued by the International Conference on Harmonization (ICH), through the paradigm of "Quality Risk Management," namely ICH Q8—Pharmaceutical Development, ICH Q9—Quality Risk Management, and ICH Q10—Pharmaceutical Quality System.

3.3 Importance of medicines in public health

Medicines are a critical part of modern healthcare for humans and animals. A medicine is any substance or component that is administered for the purposes of diagnosis, treatment, cure, mitigation, or prevention of a disease. This definition can be extended to include substances that modify physiological functions in healthy patients (such as preventing conception) and to other nondisease states (inducing anesthesia). The legal definitions of a medicine or a drug embody these concepts.

The terms medicinal substance, drug substance, and active pharmaceutical ingredient are used interchangeably by regulatory authorities and the pharmaceutical industry to describe the pharmacologically active component(s) of medicines, while medical product, drug product, and finished product are used to define the formulated drug substance that is administered to the patient or animal.

Hence, the way that medicines are controlled during development, manufacture, distribution, and use makes an important contribution to public health protection and also to ensuring that the public is neither exploited nor exposed to unacceptable risks from either prescribed or self-medication medicines. This level of control and assurance is achieved through GMP and regulation.

3.4   The role and development of pharmacopoeias

Pharmacopoeias have an important long-established role in the regulation of medicines. Their primary purpose is to set the standards for the active and inactive materials used in the preparation and manufacture of medicines. Thus, pharmacopoeial specifications and methods form the basis for the control of a large number of substances and materials used in the manufacture of medicines [10].

Pharmacopoeias were first developed at a national level and subsequently on a regional and an international basis. Examples of these three levels are the British Pharmacopoeia (BP), the European Pharmacopoeia (Ph. Eur.), and the International Pharmacopoeia (Ph. Int.). Other pharmacopoeias of importance in the international market for medicines are the US Pharmacopoeia (USP) and the Japanese Pharmacopoeia (JP).

These are similar documents, which in some cases present the same requirements and monographs albeit in a different language or format. However, it is important to understand that there are many differences between the requirements of pharmacopeias worldwide. Hence, microbiologists need to take care in including the correct pharmacopoeial requirements in regulatory submissions and in carrying out the correct tests for the market(s) where their products are authorized and sold. Unfortunately, this can lead to duplication of testing and different standards for the same attribute.

Pharmacopoeial standards are:

- objective, public standards of quality for medicines and their components;
- compliance requirements that provide the means for an independent judgement as to the overall quality of an article;
- requirements that apply throughout the shelf life of a medicinal product;
- used by a wide variety of organizations including suppliers, purchasers, manufacturers, inspectors, medicines regulators, and official and independent control laboratories.

Thus, a pharmacopoeia is an important legal component of the overall system for the control of medicines. It complements and assists the regulatory process by providing what are effectively minimum or "default" requirements for the registration and control of medicines.

Pharmacopoeias contain monographs for chemical substances, antibiotics, and various biological substances. They also cover pharmaceutical dosage forms, general monographs, standards for materials and containers, general control methods, reagents, and reference standards used in testing. For microbiologists, there are many chapters of relevance including sterility testing, endotoxin analysis, tests for microbial limits, mycoplasmas, antibiotic assays, and so on. The USP is the most comprehensive of the available texts.

3.5 Importance of inspections in the lifecycle of medicines

The inspection of critical activities during the product lifecycle of medicines is an important means of ensuring compliance with the relevant legal requirements and guidance. Consequently, nations are required to have inspection and enforcement organizations. These agencies also require investigation and enforcement powers that include seizure of "violative" products and, in respect of the manufacture and distribution of medicines, injunctions, financial penalties and restrictions, and suspensions of legal authorizations.

3.5.1 Inspection process

Inspections are carried out against the requirements set out in the relevant regulatory submissions and authorizations (e.g., marketing authorization) and the relevant GxPs. They may be organized in different ways (e.g., part of the assessment of marketing authorization applications, review of quality systems, product or process flow, facility review, compliance with specific guidance).

The authorities will normally review relevant regulatory documents and the inspection history of an operation before it is inspected. In the case of manufacturing sites, this will normally involve the submission and review of a Site Master File giving an overview of the site, its layout, management organization, and systems. Inspections are normally announced beforehand; however, it should be noted that the authorities could, if required, carry out unannounced inspections at any time.

The inspection will proceed following an agreed program during which the organization being inspected should receive initial verbal feedback. Further feedback should be provided at the end of the inspection, and this should be confirmed in the inspection report, a copy of which is sent to the inspected organization.

The inspecting authority will subsequently ask for a response to the inspection findings. If the response is inadequate and the issues are serious, then formal enforcement action by the authorities can be expected.

Inspection is a recurring process with an expected frequency of around 2 years or more often if an organization has either product quality or compliance problems.

In the United States, the FDA applies a risk-based approach to inspections with particular emphasis on Quality Systems Inspection Technique (QSIT). The FDA focuses on cGMP requirements of 4–6 different systems (e.g., QC; although they always review the companies quality system).

3.6 Role of the company regulatory affairs department

Each pharmaceutical organization must have a regulatory affairs department. This department should have a full understanding of the requirements applying to the company's products and should also act as the repository for the information provided to the regulatory authorities in support of applications and changes to the company's marketing authorization. This should include the countries where the products are marketed, "local" regulatory requirements, and the up-to-date registration status.

The regulatory affairs department should also have information about the company's authorization(s), to manufacture, where appropriate to import and to distribute the company's products. Students are advised to ensure that they know how to obtain this information and how to stay up-to-date with regulatory requirements.

3.6.1 Pharmacovigilance

Pharmacovigilance is the pharmacological science relating to the collection, detection, assessment, monitoring, and prevention of adverse effects with pharmaceutical products. The legal framework for pharmacovigilance for medicinal products in the EU/European Economic Area (EEA) is set out in a number of Directives, which describe the obligations of marketing authorization holders and the regulatory authorities. This requires them to set up a system for pharmacovigilance in order to:

- collect, collate, and evaluate information about reported and suspected adverse reactions;
- share relevant information to allow all parties involved to meet their obligations and discharge their responsibilities.

Information about drug safety is obtained from a number of sources including:

- spontaneous adverse drug reaction (ADR) reporting schemes, for example, the UK's "Yellow Card" scheme;
- clinical studies and investigation of health and diseases in wider populations;
- information from pharmaceutical companies and information published in medical literature;
- information from regulatory authorities worldwide and from morbidity and mortality databases.

Marketing authorization holders are required to operate a system to monitor and report back to the authorities on the safety of their products. This requires the collection and reporting of spontaneous safety events, and the collection and evaluation of safety data from various sources over the life of a medicine. For new products, this is defined in a pharmacovigilance plan that is part of the risk planning information that accompanies an application for a marketing authorization.

3.7   Documentation

The pharmaceutical and allied industries have a requirement to document activities for a variety of reasons, but the main one is to ensure the quality, and, thus, the safety and efficacy, of the products they sell. The pharmaceutical industry, like a number of other industries, has legal requirements relating to documentation that are supported by guidance on how to meet the requirements of the legislation. The pharmaceutical industry has adopted the principles of ISO9000 in terms of several standards that are based on a combination of quality system management and GMP. These include ISO15378, "Primary Packaging Materials," ISO14385, "Medical Devices," and the PS standards for "Pharmaceutical Packaging Materials" and "Pharmaceutical Excipients." The FDA guidance for industry document, "Quality Systems Approach to Pharmaceutical cGMPs Regulations" has implicitly adopted these principles as well. Furthermore, in Europe, legal obligations on pharmaceutical manufacturers regarding documentation are defined in the "GMP Directive" 2003/94/EC.

The most important requirement in relation to documentation is with specific batch manufacturing documentation. The documents must make it possible to trace the history of each batch. This traceability needs to be possible for a minimum defined period (at least 1 year after expiry of the batch). Where electronic data are used, data must be protected against loss or damage. This means having a validated system of data recall should such a recall of data be needed.

A document can be a procedure or a record. Examples of a procedure could be the standard operating procedure (SOP) for conducting a sterility test or the routine testing of a purified water system. A record is often related to a specific SOP and carries the confirmatory details required of that SOP. The record of the sterility test SOP could be details of the product name, its batch number, and its test result.

When designing documentation, it is helpful to think of documentation as a process. The first stage can be described as event capture, but the information or event has no status unless it can be verified or approved, which is the second stage. The last part of the process is to communicate the event, which may be by circulating and implementing a document. To illustrate this, consider the Gram stain technique. To document the procedure we need to write down the steps that capture the process. As a part of a controlled system, the steps need to be verified as being correct and the procedure "signed off" (approval stage). The procedure can be issued in to routine use along with associated training, which is the communication stage.

GMP makes certain requirements of a documentation system such as:

- assigning responsibility to an individual for control of the system;
- ensuring layout, approval, authorization, and unique identification of all documents is provided for (often by a master "documentation SOP");
- having a master "documentation SOP" to include:
  - procedures for issue, retrieval, reissue, maintenance of currency, and traceability,
  - procedures for determining the need for documents,
  - identification of documents to be included in batch dossiers (for batch release),
  - linkage of documents to licences and regulatory requirements,
  - outlining audit requirements for the documentation system,

- ensuring that only the most up-to-date version is ever used,
- retention times and archiving.

Further considerations regarding the system controlling documentation suggests that:

- documents should be available at point of use;
- master copies, including electronic versions, are held under control;
- there is control over format;
- there is a slick system for changes, approval, and reissue;
- there is control of documents of an external origin.

The majority of these requirements also make up the elements of a "documentation lifecycle."

3.7.1   Types of documents

3.7.1.1   Specifications

Specifications tend to be documents that related to starting and packaging materials, as well as finished products. They describe the standards to which these materials and product must comply if they are to be approved for use or sale. Specifications are important for microbiological testing; for example, defining pass/fail criteria. A finished product specification should contain amongst others:

- the designated name of the product and the code reference where applicable;
- the formula or a reference to;
- a description of the pharmaceutical form and package details;
- directions for sampling and testing or a reference to procedures;
- the qualitative and quantitative requirements, with the acceptance limits, for example, the sterility test or absence of specified pathogens;
- the storage conditions and any special handling precautions, where applicable;
- the shelf-life.

3.7.1.2   Instructions

All instructions to personnel (e.g., media manufacture, bacterial identification, and so on) should be clear, precise, unambiguous, and written in numbered steps, in the imperative. They should be written in a language and style that the user can readily understand. Again, detail of the content can be found in Chapter 4, particularly Sections 4.14–4.16. Associated with instructions are records and these can be either combined with the instruction or a separate document. Content of batch manufacturing records (including media batches) can be found in Sections 4.17–4.18.

3.8   Conclusion

This chapter has covered the subjects of GMP, inspection, documentation, and regulatory affairs. The object was to provide some introductory information as to how the pharmaceutical sector operates. This chapter thus complements Chapter 2 in providing the skeletal structure within which the pharmaceutical microbiologist must operate.

Of the elements covered, GMP is of overriding importance. GMP is the practices required in order to conform to guidelines recommended by agencies that control authorization and licensing for manufacture and sale of drug products, and active pharmaceutical products. These guidelines provide minimum requirements that a pharmaceutical product manufacturer must meet to assure that the products are of high quality and do not pose any risk to the consumer or public. As Chapter 1 emphasized, microbiological risks can be significant, and microbiological control is an important component of GMP.

References

[1]   Walsh G. Biopharmaceuticals: biochemistry and biotechnology. 2nd ed. Chichester, UK: Wiley; 2003, p. 78.
[2]   Baird RM, Bloomfield SF. Microbial quality assurance in cosmetics, toiletries and non-sterile pharmaceuticals. 2nd ed. Bristol, PA: Taylor & Francis; 1996, p. 51, 115.
[3]   World Health Organization (WHO). Good manufacturing practices for pharmaceutical products. In: WHO Expert Committee on specifications for pharmaceutical preparations. Thirty-seventh report. Geneva: World Health Organization; 2003 [WHO Technical Report Series, No. 908, Annex 4].
[4]   Grazal JG, Earl DS. EU and FDA GMP regulations: overview and comparison. Qual Assur J 1997;2:55–60.
[5]   Verma K. Base of a research: good clinical practice in clinical trials. J Clin Trials 2013;3:128.
[6]   EU GMP. EudraLex. Good manufacturing practice (GMP) guidelines, vol. 4. Brussels, Belgium: European Commission; 2014.
[7]   Sandle T, Saghee MR. Compliance aspects of sterile manufacturing. In: Saghee MR, editor. Achieving quality and compliance excellence in pharmaceuticals: a master class GMP guide. New Delhi: Business Horizons; 2012. p. 517–60.
[8]   Sandle T. Qualification and validation. In: Saghee MR, editor. Achieving quality and compliance excellence in pharmaceuticals: a master class GMP guide. New Delhi: Business Horizons; 2012. p. 169–206.
[9]   Sandle T, Lamba SS. Effectively incorporating quality risk management into quality systems. In: Saghee MR, editor. Achieving quality and compliance excellence in pharmaceuticals: a master class GMP guide. New Delhi: Business Horizons; 2012. p. 89–128.
[10] Akers MJ. Sterile drug product: formulation, packaging, manufacturing and quality. New York: Informa Healthcare; 2010, p. 382.

Laboratory management and design

4.1 Introduction

To produce results from microbiological analysis of good quality and to carry out such analysis in a safe and controlled way, a dedicated laboratory space is required (the activity of pharmaceutical microbiology should not be shared with other quality control activities such as analytical chemistry). The object of the quality control laboratory is to assess results from the manufacturing process and process environment to ensure that the results produced are free from test errors. To do so requires maintained equipment, ensuring that environmental controls are met, maintaining equipment, and having appropriately trained staff [1]. Furthermore, in relation to good manufacturing practice (GMP), compliance specifically all pharmaceutical quality control laboratories must be run in a compliant manner [2]. A specific area of GMP is dedicated to laboratories (in some GMP systems this is referred to as "good control laboratory practice"; this should not be confused with "good laboratory practice," which refers to animal testing facilities).

This chapter considers some of the important aspects of the design of pharmaceutical microbiology laboratories. The chapter additionally considers the key aspects of the laboratory management function. In doing so, the chapter only provides an overview of the subject, with an emphasis upon control and safety.

4.2 Pharmaceutical microbiology laboratories

Within the pharmaceutical microbiology laboratory, a range of tests are undertaken. These relate to many of the areas discussed throughout the book and invariably include bioburden testing of in-process samples and raw materials; the incubation and reading of environmental monitoring samples; water analysis; endotoxin testing; end product sterility testing; and so forth [3].

The types of microbiological tests, whether the laboratory is dedicated to sterile or to non-sterile activities, can be grouped into:

- *Quantitative examinations*: these measure the quantity of microorganisms present in the sample, and measurements need to be accurate and precise. The measurement produces a numeric value, expressed in a particular unit of measurement. For example, the number of colony forming units obtained from a bioburden sample;
- *Qualitative examinations*: tests that measure the presence or absence of microorganisms (as with the sterility test), or evaluate cellular characteristics such as morphology through microscopic examination. The results are not expressed in numerical terms, but rather in qualitative terms such as "positive" or "negative"; "reactive" or "nonreactive"; "normal" or "abnormal"; and "growth" or "no growth";

Pharmaceutical Microbiology. http://dx.doi.org/10.1016/B978-0-08-100022-9.00004-9

- *Semiquantitative examinations*: are tests that are expressed as an estimate of how much of the measured substance is present. An example here would be the gel-clot form of the limulus amebocyte lysate test.

With each of these types of tests, the laboratory results must be as accurate as possible. To support this, all aspects of the laboratory operations must be reliable and reporting needs to be timely in order for the results to be useful, especially in the event of an out-of-limits result being recorded.

4.3 Laboratory management

The manager of the microbiology laboratory should be someone of experience and qualified in microbiology. The role, depending upon the size of the institution, does not necessarily need to be the same as that of the site microbiologist.

The management function of the laboratory must be responsible for:

- establishing the policies and procedures within the quality system (including suitable standard operating procedures, SOPs);
- ensuring all policies, processes, procedures, and instructions are documented;
- making sure that all personnel understand documents, instructions, and their duties and responsibilities. This requires an efficient training and assessment system;
- providing personnel with the appropriate authority and resources to carry out their duties. This will include biological scientists and microbiologists at graduate level.

To enable these duties to be fulfilled, there needs to be a focus on time management, especially in relation to managing the time of the laboratory staff. Building a hierarchy helps the manager to deal with the workload. Supervisory staff should report to the laboratory manager, and senior scientists should support and train junior scientists.

A second area of importance is with developing the laboratory workflow. A workflow can apply to one or more tests. The sum of operations that are required for a test is called the path of workflow. The path of workflow begins with the sample and ends in reporting and results interpretation. In order to have a functioning quality management system, the structure and management of the laboratory must be organized so that quality policies can be established and implemented.

Arguably the most important laboratory resource is competent, motivated staff. Invariably success or failure depends on the knowledge and skills of the people in the laboratory, and their commitment and motivation to perform tasks as described in the job description (job descriptions should be competency based and reflect any skills needed).

4.3.1 Training

To achieve the necessary level of competence with laboratory staff, a robust and efficient training scheme is required. Training is a process to provide and develop knowledge, skills, and behaviors to meet requirements. In this context, training is linked to the job description and competency assessment and addresses identified gaps in specific tasks to be performed by the employee [4].

On-going training can be assisted through proficiency testing. It is the most commonly employed type of quality assessment, and it can be applied to many laboratory methods. In a typical proficiency test programme, challenge samples are provided at regular intervals (such as two or four times per year). For example, analysts testing samples spiked with known levels of endotoxin and attempt to verify the level, or, alternatively, identifying a cocktail of different types of microorganisms using selective agars and identification methods.

4.3.2   Quality

Laboratory management should have a focus on quality, and this will structured by a quality management system. A quality management system can be defined as the coordinated activities required to direct and to control an organization with regard to quality. This definition is used by the International Organization for Standardization (ISO) [5]. Quality management systems are also a requirement of GMP.

4.3.3   Test methods

Sampling and testing are key features of the test laboratory, and they should be described in SOPs; with SOPs in place for each test. SOPs contain step-by-step written instructions for each procedure performed in the laboratory. These instructions are essential to ensure that all procedures are performed consistently by everyone in the laboratory. An effective SOP should be:

- detailed, clear, and concise, so that staff not normally performing the procedure will be able to do so by following the SOP. All necessary details (e.g., incubation temperature requirements and precise timing instructions) should be included;
- easily understood by new personnel or students in training;
- reviewed and approved by the laboratory management;
- up to date and appropriate.

Sample management is a key part of process control, and thus, it stands as one of the essentials of a quality management system [6]. An important area of management extends to sample control. Each sample should be clearly labeled. SOPs should be extended to cover:

- a description of what samples should be stored;
- the expiry time (with many samples for microbiological testing this can only be for a few hours, such as water samples);
- the location;
- conditions for storage, such as temperature requirements;
- the system for storage organization (one method is to store samples by day of receipt or accession number);
- frequently, samples are collected outside the laboratory and must be transported for subsequent processing and testing, and thus procedures should be extended to cover this.

With test methods, arguably the most important test aspects in relation to pharmaceutical microbiology are microorganisms and culture media. If these are not controlled in a way to ensure quality, then this undermines the performance of the laboratory. Microbiology requires use of live control organisms to verify that stains, reagents and media are working correctly (this verification requires predictable reactions with microorganisms). These cultures must be carefully maintained in the form of stock and working cultures.

In relation, the quality of media used in the microbiology laboratory must be assessed as suitable for optimal and reliable results. This is in relation to both recovery of certain species and enumeration (media testing is covered elsewhere in this book).

4.3.3.1   Safety

The management role within the laboratory should not simply be focused on test compliance and results reporting. The use of chemicals and other potentially hazardous compounds separates laboratories from other types of building spaces. Protecting the health and safety of laboratory and building occupants must also be a primary concern [7].

A laboratory safety program is important in order to protect the lives of employees and visitors, to protect laboratory equipment and facilities, and to protect the environment. Hence, laboratories must be designed to maintain the health and well-being of occupants. Potentially hazardous substances used in different laboratories include chemicals, radioactive materials, and infectious biological agents [8].

The laboratory manager must ensure that there is an adequate supply of appropriate equipment, such as:

- personal protective equipment (safety glasses, laboratory coat, and gloves);
- fire extinguishers and fire blankets;
- appropriate storage and cabinets for flammable and toxic chemicals;
- eye washers and emergency shower;
- waste disposal supplies and equipment;
- first aid equipment.

Needles, broken glass, and other sharps need to be handled and disposed of appropriately to prevent risks of infection to laboratory and housekeeping staff.

To prevent or reduce incidents caused by exposure to toxic chemicals, all chemicals, including solutions and chemicals transferred from their original containers, should be labeled with their common names, concentrations, and hazards. It is crucial that chemicals are stored properly. Corrosive, toxic, and highly reactive chemicals must be stored in a well-ventilated area, and store chemicals that can ignite at room temperature must be contained in a flammables cabinet.

Laboratory-acquired infections can occur in microbiology laboratories. Aerosols are the main sources of contamination. Such microorganisms should be handled within microbiological safety cabinets (appropriate to the biohazard class of the microorganism).

A related area of infection control is with waste disposal. All waste from the laboratory must be securely bagged. Waste should be either decontaminated on site or removed by a specialist company for incineration. With biohazard 3 and 4 waste, this should always be decontaminated on site.

The laboratory manager should conduct a risk assessment with regard to the activities to be conducted within the facility. All aspects of the work (such as the use of biological agents, ionizing radiation, equipment, or harmful chemicals) must be considered for the assessment [9]. Different countries often have set procedures for conducting risk assessments, many of which are legally binding.

4.3.3.2 Laboratory information management system

An important tool for the laboratory management is an electronic data capture and sample scheduling system. A Laboratory Information Management System (LIMS) is a software-based laboratory and information management system that offers a set of features that support a laboratory's operations. The features include workflow and data tracking support, flexible architecture, and smart data exchange interfaces [10].

LIMS allows for [11]:

- the reception and log in of a sample and its associated data;
- the assignment, scheduling, and tracking of the sample and the associated analytical workload;
- the processing and quality control associated with the sample and the utilized equipment and inventory;
- the storage of data associated with the sample analysis;
- the inspection, approval, and compilation of the sample data for reporting and further analysis.

LIMS has largely replaced paper-based documentation systems.

4.3.4 Lean labs

The "lean" concept has been applied to manufacturing plants for several decades. In recent years, the idea has been applied to laboratories [11]. This is to achieve so-termed "lean labs." With the lean laboratory approach, the adoption of a generic model is unlikely to work due to the differences between laboratories (in terms of staff numbers, types of samples, working practices, types of equipment, and so on). However, the careful adaptation of lean-manufacturing techniques can deliver benefits in terms of productivity, faster testing, and it provides a structured approach to review the necessity of the samples that are processed.

Factors that can affect the performance of the laboratory include:

- *Variable work load*: many laboratories experience variations to their workload, characterized by peaks and troughs. This can lead to times of low productivity during periods when few samples are passing through the laboratory and a failure to meet sample release targets when a high number of samples are passed through the laboratory;
- *Work in progress*: this can occur when too many samples are in a state of "work in progress." Here the laboratory may not release results in a timely or efficient manner. A scenario when this can occur where laboratory technicians efficiently test samples, but they are slow in reading or reporting them.

A related situation is where there is an imbalance with trained staff. In this second scenario, a larger number of technicians may be trained with the initial handling of samples (such as preincubation steps), but a fewer number are able to complete the tests, leading to hold-ups occurring. In relation, sample throughput can be constructed in a way so that each sample is of equal priority, which removes the need for the inefficient fast tracking of samples;

- *Long and variable lead times*: this situation can occur when samples are grouped and where a test is only run when there is a preset number of samples. While this approach can save costs, if it is maintained irrespective of batch due dates, it can lead to time delays occurring;
- Fast-track systems: this describes urgent samples that can be "fast-tracked" through at the expense of other samples. If the proportion of fast-track samples increases over time, then the situation can become unmanageable;
- *Training gaps*: training gaps refers to situations when there are fewer staff available to engage in testing the variety of different samples that are presented to the laboratory. To avoid this, greater efficiencies can be harnessed through multiskilling the laboratory team.

The lean laboratory approach can address these factors by reviewing the laboratory capacity in order to level out workflow and to harness resources better, orientating resources toward peak times. To add to this, the lean laboratory approach can direct the laboratory manager to conduct a training needs analysis in order to smooth out hold-ups and prevent samples being held as "works in progress." Other aspects include examining how samples are batched together for testing and evaluating if this optimal.

These identified factors can be developed into milestones, which include:

- **(i)** reducing lead times;
- **(ii)** introducing right-first-time concepts to reduce laboratory errors;
- **(iii)** improving approval target times, to address work-in-progress;
- **(iv)** increasing technician productivity, such as number of samples processed per work session.

Another area where the lean laboratory concept can be applicable is with helping to structure a review of the types of samples going through the laboratory.

4.4   Laboratory design

The laboratory work space and facilities must be designed so that the workload can be performed without compromising the quality of work and the safety of the microbiology staff, other laboratory personnel, and visitors.

When developing a laboratory and preparing the layout, it is important to recognize the required work capacity of the laboratory, the number of staff engaged in testing, the services (electricity, water, gas) required, and the mechanisms to control inadvertent release of microorganisms to the environment as well as cross-contaminations. Sufficient space should be provided for all activities to avoid mix-ups. Suitable space should be allocated for sample receipt and processing, reference organisms, media (if necessary, with cooling), testing, and records [12].

Furthermore, the microbiology laboratory is very operator dependent, and for this reason, the design tends to be variable depending upon the array of tasks undertaken. There are, however, areas of commonality and examples of best practice. These areas are considered in Figure 4.1.

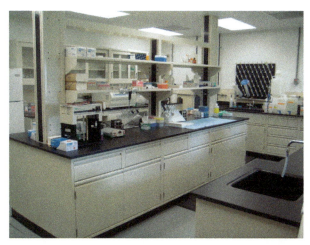

Figure 4.1 A typical microbiology laboratory.
Photograph: Tim Sandle.

4.4.1 General design

In general terms, laboratories must be fitted with large areas of bench space and storage areas. Benches must be impervious to water and solvents and must be easy to clean. All joints must be sealed, and the frame itself must be rigid and capable of supporting equipment such as safety cabinets and centrifuges. The benches must be designed to allow comfortable working. To render surfaces free of potentially infectious organisms, all surfaces will need periodic decontamination by disinfectants and, therefore, must be resistant to any such materials.

It is also important that chemically resistant vinyl flooring is used. With finishes, these must be designed and installed with the sealability of the laboratory remaining the primary consideration. There are a number of specialist materials and techniques available on the market for walls and ceilings. These include:

- vinyl cladding;
- polyvinyl chloride sheeting;
- steel panels (of the type used in the nuclear industry);
- polymer paints.

Adequate and accurate safety signage must be provided. Details of emergency contact arrangements must also be clearly displayed.

Provision must be made for sufficient laboratory coat hooks both within the lobby and the laboratory itself. Adequate storage space for clean laboratory coats must be provided within the lobby and laboratory. With microbiology laboratories, laboratory coats should not be worn outside of the department. Gloves should be worn in all instances and should be available to laboratory staff on a routine basis.

4.4.2   Sample collection and testing areas

With sample collection areas, ideally both the reception and the sample collection room are located at the entrance. This can save time and energy. With sample processing areas, this is where samples are allocated for different examinations and dispersed to the appropriate sections of the laboratory for analysis. If possible, the sample processing area should be separated from, but nearby, the testing areas.

With testing areas, ideally there will be separate rooms for the processing of environmental monitoring samples from bioburden samples, in order to avoid cross-contamination. Even where tests are carried out within the same area, measures must be taken to prevent cross-contamination of samples. Dedicated facilities should be available for sterility testing, endotoxin testing, and conducting microbial identifications.

4.4.3   Equipment

There is a variety of different types of equipment that will be required for the microbiology laboratory. These include incubators, refrigerators, microscopes, and autoclaves; such equipment must be maintained and monitored carefully. Ideally, electronic systems will measure temperature-controlled devices and alarm if temperature fluctuations outside of set parameters occur.

Considerable thought and planning should go into equipment management. The following elements should be considered:

- Selection and purchasing: when obtaining new equipment, what criteria should be used to select equipment? For example, should equipment be purchased or would it be better to lease?
- Installation: with new equipment, consider what are the installation requirements and who will install the new instrument?
- Calibration and performance evaluation: what is needed to calibrate the equipment and validate that it is operating correctly? How will these procedures be conducted for both old and new instruments?
- Maintenance: what maintenance schedule is recommended by the manufacturer? Will the laboratory need additional preventive maintenance procedures? Are current maintenance procedures being conducted properly?
- Troubleshooting: is there a clear procedure for troubleshooting for each instrument?
- Service and repair: what is the cost? Can the laboratory obtain the necessary service and repair in its geographical area?
- Retiring and disposing of equipment: what must be done to dispose of old equipment when it needs to be replaced?
- Preventive maintenance requirements. This includes measures such as systematic and routine cleaning, and adjustment and replacement of equipment parts at scheduled intervals;
- Alarm systems for equipment must be in place. To prevent the loss of samples in the event of a freezer or fridge failure.

Prior to testing samples, it is important to evaluate the performance of new equipment to ensure it is working correctly with respect to accuracy and precision. New items of equipment should be qualified. This requires similar steps, as would be applied to production equipment, to be followed: installation qualification, operational qualification, and performance qualification. A separate, but related, approach will be required for software validation.

With method validation, if the equipment or associated techniques are new, then sample validation processes will need to be considered (in relation to compendial methods, this decision will depend upon the regulatory authority). Validation can be carried out by running samples in parallel using both old and new equipment and methods for a period of time to determine that the expected results can be obtained.

One common item of equipment found in the microbiology laboratory is the autoclave. For successful sterilization to take place, a cycle must be completed at high temperature (100–150 °C) and elevated pressure (1–5 bar) in the presence of moisture, where all of the dry air has been displaced/removed, for a set period of time [13].

4.4.4   Utilities and services

The proper supply of services, such as electrical connections, gases, hot water, demineralized or distilled water, compressed air, vacuum, telephone and data networks, fire protection systems, smoke detection system and alarms, emergency showers, sprinklers, and eye-wash stations, are essential for efficient running of a laboratory.

With water, both hot and cold water pressures must be controlled so that the splashing of the users is limited. Separate wash hand basins must be available. Ideally elbow- or foot-operated systems that do not require hand touching should be used. Dispensers for soap (preferably hands free) and disposable paper towels should be permanently installed immediately adjacent to the basin(s). Due to the air disruption, and levels of contamination, hot air driers should be avoided. In addition to sinks for hand washing and sample disposal, high purity water may also be required.

For incubators and other equipments, the supply of gas (such as carbon dioxide) should be located external to the room and fed through pipework. These penetrations must be fully sealed. The need for gas monitors must be considered. Alarms from these must be audible inside and outside the laboratory.

Consideration should also be given to appropriate levels of lighting. The laboratory should provide suitable illumination; here windows are as important as fluorescent lighting. This is to allow access to natural light and for safety so that those outside can view activities inside. Another factor is sound. In providing a comfortable working environment, noise output from all equipment must be considered and adequately controlled.

4.4.5   Air supply

Three drivers determine the required volume of supply air in a laboratory: temperature, exhaust, and ventilation. Air quality control is needed for the performance of several tests; for operator comfort; and for ensuring that energy usage of efficient. With microbiological tests, some tests require set temperature and humidity ranges. For example, endotoxin testing devices can be affected by high temperatures (above 25 °C).

Consideration should be given to operator comfort. Comfort primarily is concerned with maintaining appropriate temperatures and air velocities. Worker productivity will be compromised if the space is too warm or too cool. These factors must be considered in relation to design of the air supply. Thermal comfort is of particular importance and must be carefully considered. The over-riding consideration when designing such a system must remain on simplicity.

Control of the air supply should take into account energy efficiency. Conditioning, supplying, and exhausting the large volumes of air used in laboratories consumes sizeable quantities of energy. Reducing these energy costs often forms part of the laboratory management function.

4.4.6   Clean air devices and containment

Containment laboratories must be designed and built to prevent or control the exposure of laboratory workers, other persons, and the environment to the biological agent in use. With biological agents, microorganisms are commonly placed into biohazard groups. The terms applied to the groups vary according to different regions of the world; however, they generally conform to four hazard groups (1 being lowest risk, 4 being highest) on the basis of their infectivity and the consequences of such infection [14].

For microbiology laboratories, the levels of containment usually required for work with such agents are determined by their categorization (e.g., containment Level 3 is required for Hazard Group 3 pathogens), and these reflect the increasing levels of health risk to people involved in (or who could be affected by) such work.

In order to minimize risk, most laboratories will have two physical layers of containment. The primary barrier (safety equipment), which will contain the hazard at source, and the secondary barrier (the laboratory itself), the design of which is essential in protecting both the worker and people outside the laboratory. All air leaving the containment room must pass through a HEPA (high efficiency particulate air) filter. As a further measure, laboratories are maintained at a lower pressure than surrounding areas (negative pressure), in order to prevent contaminants from spreading through a building. In constant volume laboratories, the supply and exhaust airflows are balanced to always maintain a given airflow.

Work involving hazardous microorganisms should be undertaken within a microbiological safety cabinet. Such cabinets are designed to protect the operator from the microorganism and to prevent cross-contamination of the microorganism itself. Care must be taken in positioning equipment that might generate air currents (such as fans and heaters). The safety cabinets should be installed in suitable sites in the laboratory.

Fume hoods are safety devices that are used to contain chemicals with long-term exposure hazards. Fume hoods are not appropriate for protection from substances causing significant health consequences with only isolated, short-term exposures.

4.5   Conclusion

This chapter has examined the microbiology laboratory and, in doing so, has considered laboratory management, which is based around the planning of personnel and resources, and the appropriate design of a microbiology laboratory. Design needs to be based on the objectives of safety and efficiency. Most importantly, microbiology laboratories should be designed to suit the operations to be carried out within them (as a part of the quality-by-design philosophy, in which quality is built in at the outset and not added to afterward).

Consideration should also be given to the laboratory space. Future expansion of activities, including increases in workload and staff. The design should include provision for a minimum of 25% of expansion. The structure should be flexible to allow room functional changes and allocation of new activities.

References

[1] Turnidge JD, Ferraro MJ, Jorgensen JH. Susceptibility test methods: general considerations. In: Murray PR, Baron JE, Jorgensen H, Pfaller MA, Yolken RH, editors. Manual of clinical microbiology. 8th ed. Washington, DC: American Society of Clinical Microbiology; 2003. p. 1103.

[2] Niazi SK. Handbook of pharmaceutical manufacturing formulations: sterile products, vol. 6. Boca Ranton, FL: CRC Press; 2009, p. 5.

[3] WHO. Good practices for pharmaceutical quality control laboratories. Forty fourth report. WHO technical report series, No. 957, Annex 1. Committee on specifications for pharmaceutical preparations. Geneva: World Health Organization; 2010.

[4] ISO 9001:2008. Quality management systems—requirements. Geneva: International Organization for Standardization; 2008.

[5] Wagar EA, Tamashiro L, Yasin B, Hilborne L, Bruckner DA. Patient safety in the clinical laboratory: a longitudinal analysis of specimen identification errors. Arch Pathol Lab Med 2006;130(11):1662–8.

[6] Harding AL, Brandt Byers K. Epidemiology of laboratory-associated infections. In: Fleming DO, Hunt DL, editors. Biological safety: principles and practices. Washington, DC: ASM Press; 2000. p. 35–54.

[7] Everett K, Hughes D. A guide to laboratory design. London: Butterworths; 1981, p. 128–34.

[8] Ashbrook P, Renfrew M. Safe laboratories: principles and practices for design and remodelling. Boca Raton, FL: CRC Press; 1991.

[9] Gibbon GA. A brief history of LIMS. Lab Autom Inf Manage 1996;32(1):1–5.

[10] Skobelev DO, Zaytseva TM, Kozlov AD, Perepelitsa VL, Makarova AS. Laboratory information management systems in the work of the analytic laboratory. Meas Tech 2011;53(10):1182–9.

[11] Zidel T. A lean guide to transforming healthcare: how to implement lean principles in hospitals, medical offices, clinics, and other healthcare organizations. Milwaukee, WI: Productivity Press; 2006.

[12] Barker JH, Blank CH, Steere NV. Designing a laboratory. Washington, DC: American Public Health Association; 1989.

[13] Rutala WA, Weber DJ. Disinfection and sterilization in health care facilities: what clinicians need to know. Clin Infect Dis 2004;39:702–9.

[14] Medical Research Council. Standards for containment level 3 facilities: new builds and refurbishments. 2nd ed. London: Medical Research Council; 2005.

Microbiological culture media

5.1 Introduction

Culture media is of fundamental importance for most microbiological tests: to obtain pure cultures, to grow and count microbial cells, and to cultivate and select microorganisms. Culture media remains important even with the advances with rapid microbiological methods (indeed many rapid methods continue to rely on the growth of microorganisms on a medium). Without high-quality media then the possibility of achieving accurate, reproducible, and repeatable microbiological test results is reduced [1].

A microbiological culture medium is a substance that encourages the growth, support, and survival of microorganisms. Essentially, it is a substance designed to create nutritional conditions similar to the natural environment in which the microorganism commonly survives and reproduces. Culture media contains nutrients, growth promoting factors, energy sources, buffer salts, minerals, metals, and gelling agents (for solid media) [2]. Culture media has been used by microbiologists since the nineteenth century. For the assessment of culture media, no one definitive standard exists [3]. In light of this, this chapter presents some considerations for designing the testing regime and for the selection and control of microorganisms.

5.2 Cultivation

The cultivation of microorganisms on culture media is dependent upon a number of important factors including an optimal array of nutrients, oxygen or other gases, moisture, pH, and temperature. Important nutrients include sources of carbon, nitrogen, inorganic phosphates and sulfur, trace metals, water, and vitamins. Each nutrient is, in varying combinations, a key ingredient of microbiological culture media [4]. The nutrients function as "growth factors." A growth factor is a naturally occurring substance, like an amino acid, which is capable of stimulating cellular growth, proliferation, and cellular differentiation.

5.3 A short history of culture media

The origins of microbiological culture media can be traced to the nineteenth century when the science of bacteriology was just beginning. During this pioneering time, bacteriologists attempted, with variable success, to grow microorganisms either directly using the food or material on which the microorganism had first been observed or some compound thereof. These were primarily beef-based broths of unknown and

Pharmaceutical Microbiology. http://dx.doi.org/10.1016/B978-0-08-100022-9.00005-0

variable composition [5]. Arguably, the first to cultivate microorganisms on a growth medium, with a degree of reproducibility, was the French chemist and microbiologist Louis Pasteur (1922–1985). While acting as the administrator and director of scientific studies at the École Normale (Paris), Pasteur fashioned a media of yeast, ash, candy sugar, and ammonium salts in 1860 [6]. The object was to produce a fermentation medium. This medium contained the basic requirements for microbial growth: nitrogen (ammonium salts), carbon (sugar), and vitamins (ash). In developing the media, Pasteur made some important observations: that particular chemical features of the medium can promote or impede the development of any one microorganism and that competition occurs among different microorganisms for the nutrients contained in the media, which can lead to some species outgrowing and dominating a culture.

A wider application of materials was utilized, and consequently greater success observed, when Robert Koch (1843–1910) discovered that broths based on fresh beef serum or meat extracts (so-called bouillons, the term "broth" for liquid culture medium being analogous to broth or soup) produced optimal growth [7]. Indeed, Koch's work was so groundbreaking that the cognomen "The Father of Culture Media," oft stated in many microbiological textbooks, is not misplaced.

A significant development on from the liquid medium was with solid media. In 1881, Koch demonstrated a new technique at the International Medical Congress in London, at which Pasteur is alleged to acknowledge "C'est un grand progres" [8]. Koch had recognized the difficulties of using broth media for isolation of pure cultures and had looked for solid media alternatives (this inquiry was instrumental in Koch isolating *Bacillus anthracis*, the causative agent of anthrax) for the first time, in 1882, which represented a major step-forward in disease control [9]. Initially Koch evaluated media such as coagulated egg albumen, starch paste, and an aseptically cut slice of a potato. After limited, but ultimately encouraging results, Koch developed a meat extract with added gelatine (a colorless substance derived from the collagen inside the skin and bones of animals). The resulting "nutrient gelatine" was poured onto flat glass plates, which were then inoculated and placed under a bell jar. This new plate technique could be used both to isolate pure cultures of bacteria and to subculture them either onto fresh plates or nutrient gelatine slopes in cotton-wool plugged tubes [10].

Although nutrient gelatine was a major advance, gelatine has disadvantages as a gelling agent. However, 1 year later, Koch's attempts at a nutrient medium were advanced. In 1882, Fannie Eilshemius (née Hesse) (1850–1934) suggested replacing gelatine with agar [11]. Eilshemius had been inspired by the use of agar to prepare fruit jams and jellies (agar had been used as a gelling agent in parts of Asia for centuries. Agar (or "agar–agar") is a phycocolloid water-soluble polysaccharide derived from red–purple seaweeds (the various species of *Rhodophyceae* belonging to the genus Gelidium and Gracilaria). Agar proved to be a superior gelling agent. It is prepared by treating algae with boiling water. The extract is filtered while hot, concentrated, and then dried. Agar has physical properties that could be readily adapted for bacteriology. Agar melts when heated to around 85°C, and yet when cooled it does not form a gel until it reaches 34–42°C (a physical property called hysteresis). Agar is also clearer than gelatine, and it resists digestion by bacterial enzymes. The use of agar allows the creation of a medium that can be inoculated at 40°C in its cooled molten state and yet incubated up

60 °C without melting (a useful characteristic when examining for thermopile bacteria). Typically, a 1–12% final concentration of agar is used for solidifying culture media [12].

The ability to grow bacteria on solid media was to prove a major milestone in the development of bacteriology (and agar went on to have wider application in electrophoresis, as agarose gels, and diffusion assays). The formation of bacteria on solid media led Koch to use the word "colony" to describe the pure and discrete growth [13].

A further important development for the manufacture of solid media occurred in 1887 when Julius Richard Petri (1852–1921), another worker in Koch's laboratory, was involved in modifying the flat glass plate, common to laboratories, and produced a new type of culture dish for media. This was the Petri dish. Petri used a shallow, circular glass dish with a loose-fitting cover to culture bacteria and other microorganisms, by adding gelatine-based culture media into the dish. The key design feature of the Petri dish was the use of an overhanging lid, which was in place to keep contaminants out [14]. While there is some dispute concerning whether Petri invented the "Petri dish," or whether it was in act invented earlier by Emanuel Klein, a Slovenian scientist working in England, its application in the history of microbiology is of great importance. For many years, glass dishes were used, mainly until the mid-1960s, where advances with injection moulding technology led to Petri dishes being manufactured out of clear polystyrene plastic.

Agar provides the structure for solid microbiological media, but it does not provide the nutrients necessary for bacteria to grow. For this, "growth factors" are required (essential substances that the organism is unable to synthesize from available nutrients). With the initial production of culture media, the primary nutrient sources were derived from meat (as noted by Klebs in 1871 and Nageli in 1880 [15], who were the first to record that bacteria grow well in culture media containing partially digested meat proteins or "peptones." Peptones provide a soluble and assimilable form of all the essential mineral contents of living material as well as the organic carbon and nitrogen sources). Although the meat extract used for the earliest culture media was a rich source of many of the necessary growth factors for bacteria, it was insufficient in amino-nitrogen to allow optimal growth of a range microorganisms. In 1884, Fredrick Loeffler added peptone and salt to Koch's basic meat extract formulation. The peptone he used was an enzymatic digest of meat, at the time produced in the nineteenth century as a pharmaceutical product and prescribed for nutritional disorders. This peptone added amino-nitrogen, while the salt raised the osmolarity of the medium [16].

By the 1890s, culture media had developed to form familiar to microbiologist of the twenty-first century: clearer broths; solid media in Petri dishes; and the widespread use of peptones and agar. Nonetheless, it became increasingly apparent that there was a gap in market place for mass-produced culture media. The development of commercially produced culture media originated with the meat industry, whereby hitherto discarded by-products from the manufacture of meat products were used to produce culture media. Arguably, the most prominent example of this was the German Baron Justus von Liebig who, in the nineteenth century, established a reputation as one of the fathers of modern chemistry making important contributions to research and discovery in agriculture, animal chemistry, pharmacology, and food chemistry. The Liebig Extract of Meat Company (LEMCO) was formed in 1865 to manufacture and sell

Liebig's extract of meat. The main source of meat was from the company Fray Bentos in Uruguay (where cattle were slaughtered mainly for their hides and the meat was a little-used by-product) [17].

Liebig developed his meat extract as a food source. His initial motivation was to a provide food source, stable at room temperature, for the growing malnourished poor people in central Europe (a "beef-tea" that he described as his "extraction carnis"). This became the company's most famous product: the Oxo cube. It was only later that a use for the waste product from the manufacture of Oxo cubes was found: to manufacture microbiological culture media. By 1924, the OXO Medical Division of LEMCO and the products were sold to hospitals and laboratories. The ability of OXO to provide media on a large scale was accelerated by the development of dehydrated culture media, whereby media were preserved for long periods by removing water. Low amounts of water resulted in the media powder having a low water activity that reduced the possibility of spoilage occurring. In the United States, similar movements toward mass production occurred during the war undertaken by the American Agar Company of San Diego, California, and by the Digestive Ferments Company (Difco) [18].

Since the early days of Victorian development, culture media has undergone steady development over the past 100 years. Despite better production techniques, lower contamination rates, and improved purity, the basic principles of preparing broths and agars remain the same, and the legacy of the founding parents of bacteriology continues to be of the utmost importance.

5.4   Types of culture media

There is a range of different culture media available. Different types of culture media are typically divided, based on the physical state of the media, into:

(a) liquid culture media, commonly called "broth" (Figure 5.1);
(b) solid and semi-solid culture media, commonly called "agar" (Figure 5.2).

Such media can then be further divided into such categories as growth media (designed to grow most heterotrophic microorganisms), transport media (for preserving microorganisms), enrichment media (media designed to increase the numbers of desired microorganisms), and selective growth media.

The pharmaceutical microbiology laboratory uses a range of culture media depending upon the application required [19]. The two common general medium types are nutrient agar or broth and tryptone soya agar or broth. Tryptone soya agar (equivalent to soyabean casein digest medium), in particular, is widely used for environmental monitoring. This medium is used for the isolation and cultivation of nonfastidious and fastidious microorganisms [20]. Tryptone soya broth is used for sterility testing and as a general growth broth in microbial enumeration tests, as well as used for media simulation trials [21]. For some media filling trials, vegetable peptone broth is used as a replacement as it contains no animal products.

Other media types used include fluid thioglycollate medium, used for the growth of bacteria (aerobic and anaerobic) as a part of the sterility test. Where monitoring for fungi

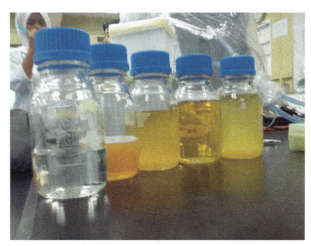

Figure 5.1 Microbiological broth media, some exhibiting turbidity.
Photograph: Tim Sandle.

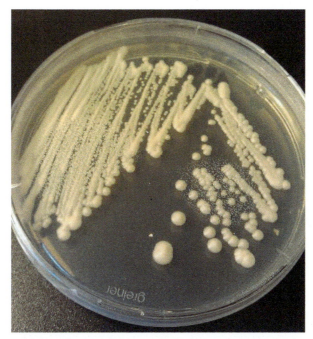

Figure 5.2 A microbiological agar plate, showing a streak plate method to obtain single colonies.
Photograph: Tim Sandle.

is required, such as part of an environmental monitoring regime, the commonly used media are sabouraud dextrose agar or malt extra agar. For the microbiological examination of water, R2A is used. This is a low nutrient agar used for the cultivation of heterotrophic microorganisms. Other media are used for microbiological identification, such as Columbia blood agar (for the detection of hemolytic reactions by Staphylococci).

The manufacture of media either manufactured "in-house" (whereby a dehydrated formulation is used) or, more commonly, purchased ready-made [22]. Where media is purchased ready-to-use, the microbiologist has a responsibility to audit the manufacturer of the media [23]. For certain plate media, such as those used in cleanrooms, the media should be sterilized by irradiation [24].

5.5   Quality control of culture media

The European and US pharmacopoeias describe the application of culture media for several tests, primarily sterility, microbial enumeration, and the examination of pharmaceutical grade water [3]. There is, however, no excepted international standard for the testing of culture media applicable to the pharmaceutical industry. There is a standard, divided into two parts, written for the food industry: ISO11133 [25,26] but one that is often drawn upon by pharmaceutical microbiology. Part one is concerned with the general terminology related to quality assurance and specifies the minimum requirements for the preparation of culture media to be used for the microbiological analysis; and part two relates to the criteria and methods for the performance testing of culture media. The standard is of further importance given that aspects of it have formed the basis of accepted testing regimes recommended by culture media manufacturers.

It is important that each batch of such media undergoes some form of quality control before it is released for general use to provide a measure of confidence that the results issued from microbiology laboratories are accurate. Testing is normally undertaken once all preparatory steps have been completed, including irradiation.

The quality control of culture media can be divided into two parts: physical characteristics and microbial characteristics.

5.5.1   Physical characteristics

The tests undertaken for the physical characteristics of culture media vary depending upon the type of media. Examples of physical tests include:

- Visual test for color: the color of a sterilized medium should be compared with a non-sterilized medium and any differences in color noted;
- Visual test for clarity: the clarity of the media should be examined for optical artefacts, such as crystallization;
- Gel-strength: the gel strength should not be over hard or over soft, but firm and usable;
- pH of the finished media: this is probably the most important chemical test, for if a pH is outside of the recommended range for the media then this will lead to the inhibition of some of the microorganisms that the media is intended to grow [27];
- Checks for damage: plates and bottles should be examined for damage like cracks and defects.

5.5.2 Microbiological characteristics

(a) The test of media sterility is designed to detect microbial contamination during the manufacturing process. Here a small number, normally 2% of the batch, of uninoculated items are incubated. The temperature and time selected for the sterility test incubation will depend upon the type of media. For general purpose media, a temperature of 30–35 °C for 3 days is typical. To pass the sterility test, the items must demonstrate no growth.

(b) Arguably the challenging of culture media with microorganisms is the most important test carried out in the microbiology laboratory. That such a key test is undertaken by the media manufacturer is unquestionable. Additionally, it is common for the purchaser to carry out growth promotion, to check for batch-to-batch variability or to assess any issues during shipment [28].

For growth promotion, a panel of microorganisms is required to demonstrate the suitability of the media for its intended use. Where the pharmacopeia recommends certain microorganisms and that these must be traceable to a reputable culture collection, such as the American Type Culture Collection, ATCC (although the pharmacopeia allows for alternative culture collections to be used there is some ambiguity about strain equivalency). Type cultures should be carefully preserved within the culture collection of the laboratory. This includes ensuring that cultures are held at a temperature low enough to avoid phenotypic variations from occurring and restricting the number of passages between subculture steps to fewer than five [29].

The standard set of typed cultures detailed in the European Pharmacopeia and United States Pharmacopeia are shown in Table 5.1.

These microorganisms have been serially subcultured in national culture collections over decades and are conditioned to growth on rich laboratory culture media. They are designed to allow the vendor to assess the media as suitable at the point of manufacture and for the user to verify the media upon receipt.

In addition to type cultures, environmental isolates are commonly used in media testing regimes. These organisms are designed to demonstrate that a particular lot of culture media will grow microorganisms that are representative of the types that are found in the manufacturing environment [30]. Thus, media used for the examination of water would have a test panel that included microbial isolates from water (such as

Table 5.1 **Standard media growth promotion test microorganisms**

Microorganism	Culture collection reference
Staphylococcus aureus subsp. *aureus*	ATCC 6538
Bacillus subtilis subsp. *spizizenii*	ATCC 6633
Pseudomonas aeruginosa	ATCC 9027
Clostridium sporogenes	ATCC 19404
Candida albicans	ATCC 10231
Aspergillus brasilensis	ATCC 16404
Escherichia coli	ATCC 8739
Salmonella enterica subsp. *enterica* serovar *typhimurium*	ATCC 13311

Pseudomonad-related bacteria), and media used for environmental monitoring would include bacteria transient to human skin (such as Staphylococci).

While the use of such isolates is increasingly becoming a regulatory expectation, the adoption of environmental or plant isolates is not supported by all microbiologists. Arguments for the use of such isolates are that the media are challenged with those microorganisms actually encountered within the pharmaceutical environment, and that these are often more representative than the standard cultures. Moreover, the isolates can be varied over time, based on reviews of microflora, so that they remain so perpetually relevant. Arguments against include the fact that interlaboratory assessments are rendered difficult because each laboratory is using a different organism set. A second point is that once organisms are grown on standard media they become indistinguishable from other laboratory strains. It has been counter argued that minimally subcultured environmental isolates have aspects of their "wildtype" attributes conserved. The outcome of this debate is ongoing, and clearly further study is needed.

5.5.3   Test methods and acceptance criteria

The numerical level of the microbial challenge is another important consideration. Most testing regimes require a low-level challenge. This is to show that the media can recover low numbers of microorganisms. In most, this is a challenge of fewer than 100 microorganisms [31].

5.5.4   Solid media

There are various qualitative and quantitative approaches that can be taken for the testing regime. For the testing of agar, qualitative approaches include simple subculture streaks (spread plates). Here, liquid cultures are streaked with an inoculation loop to give single colonies. Each segment of the agar plate can then be compared to the growth characteristics of a suitable control plate (a control medium is a released batch of media, which has previously been assessed as having good growth promotion properties). A more robust system is ensured through quantitative techniques. These generally fall into two groups: the ecometric and the Miles–Misra [32]. Both of these tests compare one set of media (a previously released batch) against another (the batch to be tested). The ecometric method is a semiquantitative variant of the streaking method [33]. One loopful of inoculm is placed onto the plate and is sequentially diluted streak to streak. Five streaks are streaked out into four quadrants onto the agar plate along with a final streak in the center of the plate. Growth should occur in all streaks.

The Miles–Misra technique (the drop count technique) involves spreading droplets of known quantities of microbial suspensions (typically $10\,\mu L$). The test plate is compared with a control plate, after incubation, in terms of the number of colonies recovered. The accuracy of the method is dependent upon the dilution used, the number of colony forming units (cfu) in the inoculum, the volume of the inoculums used, and the spreading technique [34]. The result is typically expressed as a productivity ratio when, after incubation, the count of a previously released batch of media is divided into the count of the test media [35]. This is calculated using the following formula, based on the two duplicate samples for the test plate and control plate:

$$\text{Productivity ratio} = \frac{\text{Mean of two test plates}(c\,fu)}{\text{Mean of two comparative control plates}(c\,fu)}$$

For example:

Test plates:	32 and 40 cfu (mean 36 cfu)
Control plates:	50 and 46 cfu (mean 48 cfu)
Productivity ratio:	36/48 = 0.8

An acceptable productivity ratio must be equal to or greater than 0.5 and with an upper limit of 2.0 (this is equivalent to a 50–200% recovery).

5.5.5 Broth media

For broth (liquid) media, it is more difficult to apply a quantitative assessment. Many laboratories challenge broth media with an estimated number of microorganisms and compare the growth, over time, with a control batch (which provides a qualitative assessment of copious growth). The challenge is typically fewer than 100 cfu, and the time to obtain growth is between 3 and 5 days. The growth between the test batch and the control batch is then compared with the requirement that both must show copious growth. Alternatively, some laboratories attempt a semiquantitative approach by constructing a growth index from slight to copious growth (normally a scale of +, ++, or +++).

5.5.6 Test regime

Once decisions relating to the type of microorganisms and the test method have been made, the question of the test conditions arises. Many laboratories use general media that may be used at a range of temperatures, yet to test this media at every temperature that it could potentially be used could be expensive and could create an unwieldy release system. A practical approach is to test at the mid-range temperature. However, any regime will need to be defensible to regulatory authorities.

Incubation time is another parameter that requires careful planning. For some media, this is clearly defined in the pharmacopoeias (typically growth of bacteria within three days and growth of fungi within 5 days). However, for other media, a realistic time must be established based on the application of the media and the types of microorganisms that are used for the challenge.

5.5.7 Testing of selective media

Selective media requires a slightly different approach in that the aim is not only to see if the media supports the growth of a range of microorganisms, but also to examine inhibition and colony morphology and pigment. Here positive and negative control strains are required. In order to ascertain growth characteristics for selective media, some microorganisms are used as positive or negative indicators of growth. For the positive reaction, particular colony morphologies, pigmentation, or diffusion of activity may be part of the acceptance criteria.

5.5.8   Expiry time assessment of culture media

Culture media will have defined storage conditions and expiry time and the shelf-life needs to be validated [36]. This is to assess if different humidity levels (which can affect the water activity of solid media), chemo-oxidation (due to physical factors such as heat) and photo-oxidation (from sunlight) affect the media [37].

5.6   Manufacture of culture media

The manufacture of media is an important process for a microbiology laboratory. The manufacture of media either manufactured "in-house" (within the microbiology laboratory of the pharmaceutical company whereby a dehydrated formulation is used) or, more commonly, purchased from the manufacturer ready-made or from a contract manufacturing laboratory (this media can be for immediate use, such as agar in a Petri dish or media that is partially completed and requires an interim preparatory step prior to use, such as melting). Even where media are purchased ready-to-use, the microbiologist has a responsibility to audit the manufacturer of the media and to have an understanding of the media manufacturing process.

In setting up a media preparation area (sometimes referred to, especially in North America, as the media kitchen), it is important for the microbiologist to build in quality to the design of the workflow and the key manufacturing steps. Important manufacturing concepts, such as batch rotation, for example, should apply to the use of media powder. As with other pharmaceutical manufacturing processes, aseptic techniques should be adopted at all times to minimize the opportunity for the media to become contaminated.

The key stages for media manufacture are [23]: initial preparation, rehydration, sterilization, addition of supplements, filling, labeling, and secondary sterilization.

5.6.1   Initial preparation

Prior to manufacturing, the types of media and quantities required should be planned out. When complex media are required purchasing, a premixed powder is the most straightforward option (powdered media in this form is supplied by several different companies). Ready-to-use powder is not available for all types of media, and with such cases, various individual constituents will need to be purchased.

Consideration should be given to the appropriate vessel for holding or dispensing the media. For the preparation of broth, it is normal to dispense the broth into the required in-use containers (commonly glass bottles) prior to sterilization. The use of an accurate dispensing device is required for this. A regular check of the volume dispensed, by measuring volume or more commonly by verifying the weight, is an important quality control step. For plate media, it is more common for the media preparation to be sterilized, then cooled, and then dispensed into Petri dishes. The fill volume of the plates should be confirmed during dispensing.

Some laboratories elect to perform microbiological testing on the received dehydrate powder (where there is a particular risk from spore-forming microorganisms that might, in sufficiently high numbers, survive the sterilization step). This is a separate activity to the more universal postmanufacture quality control testing of the prepared culture media.

5.6.2   Rehydration

The preparation of a culture medium requires the components to be dissolved. Media powder is rehydrated by mixing a measured amount of the medium in the required volume of water. The water should be freshly prepared and, depending on the type of media, held at a warm temperature. Instructions for rehydration are usually printed on the container (e.g., 30 g/L for agar X into 5 L of water). When undertaking this step, it is important to maintain the homogeneity of the solution by mixing. With media containing agar as a solidifying agent, the media are hydrated by gently heating and agitating the water: media mixture to dissolve it. Care must be taken to avoid scorching the media. Here, the media should clarify near boiling (95–100 °C), and the media should only be allowed to boil for a brief period of time (less than 1 min).

5.6.3   Sterilization

The sterilization of the media plays an important role in the quality of the media. Most incoming powdered media contain a level of contamination, and if the media is incorrectly sterilized (to eliminate viable microorganisms), then spoilage could develop thereby rendering the media unusable. Most media are sterilized by steam under pressure (autoclaving) or, for some plate media, in an agar preparator. For some special culture, media sterilization is performed by filtration (where heat would destroy one or more growth factors). For all media and reagents that are sterilized by autoclaving, a sterilization batch number should be assigned. The procedure for assigning the sterilization batch number should follow a simple and similar notation. A suitably qualified autoclave should be used.

For containers holding agar media prior to dispensing, the container should be held at around 50 °C in preparation for filling. The medium should be dispensed as soon as it equilibrates to 45–50 °C or within a maximum time (3 h is a typical target time). It is recommended that the medium is mixed gently prior to dispensing.

5.6.4   Addition of supplements

For some types of media, additives or supplements are required. Most of these components cannot be sterilized by autoclaving as they are heat labile, such as buffers and amino acids. For these materials, the standard practice is to sterilize them by membrane filtration (the removal of microorganisms from the liquid). The type of filter used will depend upon the solution type. Care must be taken when selecting the material of the filter to avoid any leechables contaminating the filtrate. Such supplements need to be added to the media after sterilization, when the media are held at a cooler temperature in the water bath.

5.6.5 Filling

For agar, the sterilized media require dispensing into Petri dishes. This activity requires utmost caution to avoid contamination. To prepare filled Petri dishes, the sterilized media should be carefully tempered while in the molten state (to around 45–50 °C, gelling typically occurs between 32 and 40 °C) and then dispensed into sterile, glass, or plastic Petri dishes. Petri dishes are typically of a 95–100 mm diameter or to a 50–55 mm diameter (for contact plates used for surface sampling as parts of an environmental monitoring regime) and are purchased in sleeves of dishes (often with 20 or 50 plates per sleeve). Agar does not normally distribute uniformly when melted and requires mixing to ensure a uniform distribution. The filling activity is normally a semiautomated process using dish fillers. Periodic assessments of the volume filled must take place during the filling operation. This check is normally made by weighing a selection of filled plates. The dispensing of plates should take place under a unidirectional airflow hood to minimize contamination. This is necessary to avoid contamination even if the plates are to be subject to a secondary sterilization step such as irradiation.

5.6.6 Labeling

Once the sterilization and filling steps have been competed, all sterilized broth bottles and filled plates must be labeled, and the details recorded in the batch record. Labels should specify the media type, batch number, expiry time, and storage conditions.

5.6.7 Secondary sterilization

For certain plate media, such as that used in cleanrooms, an additional sterilization step can be undertaken. The most common method for this is irradiation by ionization using gamma rays. The effectiveness of the irradiation should also be checked by a sterility test of the media (which is discussed below). The sterility check is important given that the application of the irradiation dose and the combination of different loading patterns can affect the process [20].

5.7 Media release and quarantine

A media quality control system will need to consider the release criteria and quarantine system. With regard to the release criteria, the laboratory must put in place clear guidelines for the repeat test procedure. This will need to cover invalid tests and the procedure to be followed should any microorganisms fail to grow or show recovery at the expected level [38]. A quarantine system is important in order to prevent media that have not been assessed from entering general use.

5.8 Summary

Microbiological culture media are the most widely used and arguably most important "tool" of the pharmaceutical microbiologist. Given this primacy, it is important that the media manufactured or purchased by the laboratory are of high quality and suitable for the intended test method. This chapter has set out to show that the control and release of microbiological culture media require a well thought-out structure. The chapter has also considered some of the important practices to observe in relation to the manufacture of culture media. Given the importance of culture media in relation to different microbiologists' tests, the importance that should be paid to media manufacture, control, testing and release cannot be underestimated: culture media is the bedrock of most microbiological examinations.

References

[1] Sandle T. Selection of microbiological culture media and testing regimes. In: Saghee MR, Sandle T, Tidswell EC, editors. Microbiology and sterility assurance in pharmaceuticals and medical devices. New Delhi: Business Horizons; 2010. p. 101–20.

[2] Bridson E, Brecker A. Design and formulation of microbiological culture media. In: Noris JR, Ribbons DW, editors. Methods in microbiology, vol. 3A. London: Academic Press; 1970. p. 229–95.

[3] Cundell A. Review of media selection and incubation conditions for the compendial sterility and microbial limits tests. Pharm Forum 2002;28(6):2034–41.

[4] Stainer RY, Ingraham JL, Wheelis ML, Painter PR. General microbiology. 5th ed. Basingstoke: Macmillan; 1987, p. 22–3.

[5] Tseng CK. Agar. In: Alexander J, editor. Colloid chemistry, vol. 6. New York: Reinhold Publishing; 1946. p. 629.

[6] Collard PJ. The development of microbiology. Cambridge: Cambridge University Press; 1976, p. 25–8.

[7] Koch R. Untersuchungen ueber Bakterien V. Die Aetiologie der Milzbrand-Krankheit, begruendent auf die Entwicklungsgeschichte des Bacillus Anthracis. Beitr z Biol D Pflanzen 1876;2:277–310 [In Milestones in microbiology: 1556 to 1940, translated and edited by Thomas D. Brock, ASM Press; 1998. p. 89].

[8] Sakula A. Baroness Burdett-Coutts' garden party: the International; Medical Congress, London, 1881. Med Hist 1982;26:183–90.

[9] Koch R. Die Aetiologie der Tuberculose. Berl Klin Wchnschr 1882;xix:221–30 [In Milestones in microbiology: 1556 to 1940, translated and edited by Thomas D. Brock, ASM Press; 1998. p. 109].

[10] Koch R. Zur Untersuchung von pathogenen Organismen. Mitth a d Kaiserl Gesundheitsampte 1881;1:1–48 [Cited in Milestones in microbiology: 1556 to 1940, translated and edited by Thomas D. Brock, ASM Press; 1998. p. 101].

[11] Bridsen E. The development, manufacture and control of microbiological culture media. Basingstoke, UK: Oxoid; 1994.

[12] Sandle T. History and development of microbiological culture media. J Inst Sci Technol 2011;(Winter):10–4.

[13] Mortimer P. Koch's colonies and the culinary contribution of Fanny Hesse. Microbiol Today 2001;28:136–7.

[14] Hitchens AP, Leikind MC. The introduction of agar-agar into bacteriology. J Bacteriol 1939;37:485–93.

[15] Merck E. Bericht uber das Jahr 1982. p. 84 [Cited in Metz, H. Culture media: then and now. Med Technol; March 1990. p. 14–5].

[16] Loeffler F. Mittheil Kaiserl Gesunheitsante 1884;2 [Cited in Brock, T. Robert Koch: a life in medicine and bacteriology. Science Technical Publications. Madison, WI; 1998].

[17] Graham-Yool A. The forgotten colony: a history of the English speaking communities in Argentina. London: Hutchinson; 1981.

[18] Robertson GR. The agar industry in California. Ind Eng Chem 1930;22(10):1074–107.

[19] Sutton SVW. Activities of the USP analytical microbiology expert committee during the 2000–2005 revision cycle. J Pharm Sci Technol 2005;59(3):157–76.

[20] Barry AL, Fay GD. A review of some common sources of error in the preparation of agar media. Am J Med Technol 1972;38:241–5.

[21] Baird RM, Corry JEL, Curtis GDW. Pharmacopoeia of culture media for food microbiology. London: Elsevier Science; 1986.

[22] Sandle T. Selection and use of cleaning and disinfection agents in pharmaceutical manufacturing. In: Hodges N, Hanlon G, editors. Industrial pharmaceutical microbiology standards and controls. England: Euromed Communications; 2003. p. 7.1–36 [Chapter revised on several occasions].

[23] Sandle T. The media kitchen: preparation and testing of microbiological culture media. In: Sutton S, editor. Laboratory design: establishing the facility and management structure. Bethesda, MD: Parenteral Drug Association; 2010. p. 269–93.

[24] Booth C. Media fills—trial or triumph. Lab News 2006;16–7.

[25] ISO/TS 11133-1. Microbiology of food and animal feeding stuffs—guidelines on preparation and production of culture media: part 1: general guidelines on quality assurance for the preparation of culture media in the laboratory; 2009.

[26] ISO/TS 11133-2. Microbiology of food and animal feeding stuffs—guidelines on preparation and production of culture media: part 2: practical guidelines on performance testing of culture media; 2003.

[27] Evans GL, Bell RH, Cunningham LV, Ferraro MJ, Maltese AE, Pienta PA. Quality assurance for commercially prepared microbiological culture media: approved standards. 2nd ed. USA: National Committee for Clinical Laboratory Standards; 1996, M22-A2, p. 16.

[28] Nagel JG, Kunz LJ. Needless retesting of quality-assured commercially prepared culture media. Appl Microbiol 1973;26(1):31–7.

[29] Snell JJS. Preservation of control strains. In: Snell JJS, Brown DFB, Roberts C, editors. Quality assurance: principles and practice in the microbiology laboratory. London: Public Health Laboratory Service; 1995. p. 69–76.

[30] Brown MRW, Gilbert P. Microbiological quality assurance: a guide towards relevance and reproducibility of inocula. USA: CRC Press; 1995.

[31] Baird RM, Corry JEL, Curtis GDW. Pharmacopoeia of culture media for food microbiology. In: Baird RM, Corry JEL, Curtis GDW, editors. Proceedings of the 4th international symposium on quality assurance and quality control of microbiological culture media, Manchester, 4–5 September. Int J Food Microbiol 1989; 5:187–299.

[32] Mossel DAA, et al. Quality control of solid culture media: a comparison of the classic and the so-called ecometric technique. J Appl Bacteriol 1980;49:439–54.

[33] Mossel DAA, Bonants-van Laarhoven TMG, Lichtenberg-Merkus AMT, Werdler MEM. Quality assurance of selective culture media for bacteria, moulds and yeasts: an attempt at standardisation at the international level. J Appl Bacteriol 1983;54:313–27.

[34] Martin R. Culture media in quality control: principles and practice in the microbiology laboratory. London: PHLS; 1991.

[35] Mossel D. Introduction and perspective. Int J Food Microbiol 1985;2:1–7.

[36] Vanderzantz C, Splittstoesser DF, editors. Compendium of methods for the microbiological examination of foods. 3rd ed. USA: American Public Health Association; 1992. p. 92.

[37] Nichols E. Quality control of culture media. In: Snell JJS, Brown DFB, Roberts C, editors. Quality assurance: principles and practice in the microbiology laboratory. London: Public Health Laboratory Service; 1989. p. 119–50.

[38] Sutton S. Microbial recovery studies—50% or 70%? Pharm Microbiol Forum Newsl 2007;13(7):3–9.

Microbiology laboratory techniques

6

6.1 Introduction

In the manufacture of all types of pharmaceuticals, quality assurance represents a major consideration. It is important that products are not contaminated with microorganisms that might affect their safety, efficacy, or acceptability to the patient.

During pharmaceutical product manufacture, microbiological contamination is controlled by the application of good manufacturing practice (GMP). Nonetheless, contamination risks remain an ever-present threat. In practice, the presence of microorganisms in pharmaceutical products constitutes two main hazards.

(1) It could result in spoilage of the product; the metabolic versatility of microorganisms is such that any formulation ingredient may undergo degradation in the presence of a given microorganism.

(2) It may provide an infection hazard to the patient. Although the degree of hazard will be dependent on the product's intended use and route of administration (i.e., oral, topical, parenteral, application to the eye, and so on), With nonsterile products, certain pathogens present a hazard; with sterile products, any contamination presents a potential risk.

For these reasons, the microbiological contamination control of pharmaceuticals is evaluated during various phases of product development and during routine commercial manufacture [1]. This requires a range of microbiological tests to be conducted. Such tests focus on the number and type of microorganisms (and any potential microbial impurities such as bacterial endotoxins). These tests are applied to raw material ingredients (including pharmaceutical waters), in-process product checks (including bioburden levels) during manufacture, environmental controls and ultimately finished product microbiological tests. Many of these tests are described in the various national and international pharmacopeia. To add to these, there are techniques to assess cleanrooms through the environmental monitoring program.

Microbiology plays a critical role in pharmaceutical quality control, specifically evaluating raw materials, process controls, product release tests, and product stability tests. The quality and interpretation of the data from these tests critically impacts product safety. It is the quality control function that assures that data from these tests are meaningful (reliable and precise) and have a minimum of error [2]. Microbiologists must evaluate the suitability for the use of microbiology tests, the limitations of their applicability and measurements, and whether acceptance criteria were met. Furthermore, microbiologists must understand both the nature of the tests and the data derived from them.

Pharmaceutical Microbiology. http://dx.doi.org/10.1016/B978-0-08-100022-9.00006-2

This chapter describes some of the more commonly performed tests and focuses on those tests that are not described elsewhere in dedicated chapters.

6.2   Good laboratory practice and laboratory safety

The use of good laboratory practice is an important factor in safeguarding the health and safety of laboratory personnel. It should be remembered that many of the bacteria that are cultured in microbiological laboratories are capable of producing disease in humans. This, coupled with the fact that, potentially more virulent, pure strains of such bacteria are often being produced, means that there is considerable risk to the health of microbiology laboratory workers if adequate precautions are not taken.

The basis of good practice in a microbiological laboratory can be summed up by the following:

- ensure all necessary equipment and media are sterilized prior to use;
- ensure that all sterilized equipment and media is not re-contaminated after sterilization by allowing it to touch, or rest on, any unsterilized surface;
- frequently disinfect hands and working surfaces;
- as far as possible, eliminate flies and other insects that can contaminate surfaces, equipment, media, and also pass organisms to laboratory personnel;
- never pipette by mouth samples that are suspected to have high bacterial concentrations;
- wear appropriate protective clothing: laboratory coat, safety glasses, and gloves;
- do not eat, drink, or smoke in the laboratory;
- sterilize contaminated waste materials prior to disposal;
- take care to avoid operations that result in bacterial aerosols being formed.

Each laboratory should have a risk assessment system in place. This is based on possible hazards and the risks associated with them. Taking microorganisms as an example, here a hazard is the danger or harm that a microorganism may cause to a person. A risk is the probability or likelihood that a person will be harmed by the microorganism. Safety issues, including protective clothing, are considered in Chapter 4.

6.3   Aseptic technique

Due to the fact that microorganisms can be present virtually anywhere, it is important to take measures to avoid contamination of microbiological experiments with extraneous bacteria. The measures used to prevent this cross-contamination in microbiological laboratories are collectively known as aseptic techniques.

Aseptic techniques usually involve disinfection of working areas, minimizing possible access by bacteria from the air to exposed media and use of flames to kill bacteria that might enter vessels as they are opened.

Asepsis can be achieved by laboratory staff washing their hands and disinfecting the bench area. Work should be conducted in a dust-free and draught-free area, using a unidirectional airflow cabinet for critical activities. In terms of the application of techniques, staff should not touch any part of the container, pipette, and so, that will come in contact with the sample or culture.

6.4 Cultures and identifications

An important aspect of microbiology is with cultural techniques and in obtaining a pure culture. Microbiological culture describes a method of multiplying microbial organisms by allowing them to reproduce in predetermined culture media under controlled laboratory conditions (time, temperature, and atmospheric conditions). Microbial cultures are used to determine the type of organism, its abundance in the sample being tested, or both.

Furthermore, microbial cultures are foundational, and they are required for basic diagnostic methods. For these, it is necessary to isolate a pure culture of microorganisms. A pure (or axenic) culture is a population of cells or multicellular organisms growing in the absence of other species or types. A pure culture may originate from a single cell or single organism, in which case the cells are genetic clones of one another.

Developing pure culture techniques is crucial to the observation of the specimen in question. The most common method to isolate individual cells and produce a pure culture is to prepare a streak plate. This method is a means to separate the microbial population physically and is performed by spreading and then inoculating back and forth with an inoculating loop over the solid agar plate. Upon incubation, colonies will arise, and single cells will have been isolated from the biomass.

In essence, the steps required are, for transfer onto solid agar:

- Sterilize a wire loop (or use a sterile plastic disposable loop) by heating it until red hot in a flame; allow it to cool for several seconds. Test for coolness by touching the agar at the edge of the plate;
- Pick up a loop full of liquid inoculum or bacterial growth from the surface of an agar plate and, starting about 2.5 cm in from the edge of the plate, streak lightly back and forth with the loop flat, making close, parallel streaks back to the edge of the plate;
- Sterilize the loop and cool again, then with the edge of the loop, lightly make another set of nearly parallel streaks about 0.32 cm apart, in one direction only, from the inoculated area to one side of the uninoculated area, so that about one-half of the plate is now covered;
- Flame and cool the loop again, and make another set of streaks in one direction, perpendicular to and crossing the second set of streaks, but avoiding the first set.

The pure culture is a foundation method for conducting microbial identifications, as described in Chapter 9.

6.5 Microscopy

The light microscope is an important tool in the study of microorganisms, particularly for identification purposes. The compound light microscope uses visible light to directly illuminate specimens in a two-lens system, resulting in the illuminated specimen appearing dark against a bright background. The two lenses present in a compound microscope are the ocular lens in the eyepiece and the objective lens located in the revolving nosepiece. Compound light microscopes typically have the following components (as outlined below and set out in Figure 6.1):

• Illuminator: the light source in the base of the microscope;
• Abbe Condensor: a two lens system that collects and concentrates light from the illuminator and directs it to the iris diaphragm;
• Iris diaphragm: regulates the amount of light entering the lens system;
• Mechanical stage: a platform used to place the slide on which has a hole in the center to let light from the illuminator pass through. Often contains stage clips to hold the slide in place;
• Body tube: houses the lens system that magnifies the specimens;
• Upper end of body tube—oculars/eye pieces: what you view through;
• Lower end of body tube—nose-piece: revolves and contains the objectives.

Essentially, a light microscope magnifies small objects and makes them visible. The science of microscopy is based on the following concepts and principles:

• Magnification is simply the enlargement of the specimen. In a compound lens system, each lens sequentially enlarges or magnifies the specimen;
• The objective lens magnifies the specimen, producing a real image that is then magnified by the ocular lens resulting in the final image;
• The total magnification can be calculated by multiplying the objective lens value by the ocular lens value.

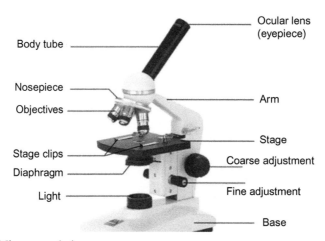

Figure 6.1 Microscope design.

6.6 Pharmacopeia and microbiological tests

The majority of tests to assure the microbiological quality assurance of pharmaceutical products are described in the major pharmacopoeias (such as the British Pharmacopoeia (BP), European Pharmacopoeia (Ph. Eur.), US Pharmacopoeia (USP), Japanese Pharmacopoeia (JP), and the World Health Organization International Pharmacopeia). Of these, the Ph. Eur, USP, and JP constitute the primary texts. The tests described constitute the set of basic microbiological laboratory techniques in relation to pharmaceuticals and healthcare [3]. Alternative tests to the pharmacopeia described methods can be validated and employed [4], but the pharmacopoeial method remains the referee test in the event of any dispute over product quality (this is the case, for example, with many of the types of rapid methods described in Chapter 17).

The basic methods are shown in Table 6.1 with some additional supporting documents that provide guidance on the microbiological quality expectations of pharmaceutical preparations and good microbiological laboratory practice.

Table 6.1 **Primary pharmacopoeial microbiology tests**

Pharmacopoeial chapter/section	Application
Microbiological examination of nonsterile products: total viable aerobic count (Ph. Eur. 2.6.12, USP <61>) USP <2021> Microbial enumeration tests-nutritional and dietary supplements USP <2023> Microbiological attributes of nonsterile nutritional and dietary supplements	Number of organism in raw materials, water, product in-process controls (bioburden), finished products
Microbiological examination of nonsterile products: tests for specified organisms (Ph. Eur. 2.6.13, USP <62>) USP <2022> Microbiological procedures for absence of specified microorganisms-nutritional and dietary supplements	Type of organisms present in raw materials, water, product IPC (bioburden), finished products
Ph. Eur. 5.1.4 Microbiological quality of pharmaceutical preparations/USP <1111> Microbiological attributes of nonsterile pharmaceutical products	Setting of limits and control factors
Sterility (Ph. Eur. 2.6.1, USP <71>)	Sterility test for finished products
Pyrogens/endotoxin Rabbit Pyrogen Test (Ph. Eur. 2.6.8, USP<151>) *Limulus* amoebocyte lysate (LAL) bacterial endotoxin test (Ph. Eur. 2.6.14, USP<85>)	Pyrogen/endotoxin test for raw materials, pharmaceuticals waters, product IPC, finished products

Continued

Table 6.1 **Continued**

Pharmacopoeial chapter/section	Application
Antimicrobial preservative efficacy testing (Ph. Eur. 5.1.3, USP <51>)	Product formulation challenge to microbiological contamination
Microbiological assay of antibiotics (E.P 2.7.2., USP<81>)	Potency assays for antibiotic pharmaceutical preparations
USP <1112> Application of water activity determination to nonsterile pharmaceutical products	Assessment of water activity (can affect microbial growth and survival)
USP <1211> Sterilization and sterility assurance of compendial articles. There are a series of subchapters that describe specific sterilization methods	Sterilization and microbial kill
USP <55> Biological indicators	Biological indicators for assessing microbial kill
USP <1113> Microbial characterization, identification, and strain typing	Microbial identification methods
USP <1117> Microbiological best laboratory practice	General laboratory methods and management
USP <1116> Microbiological control and monitoring of aseptic processing environments	Environmental monitoring and cleanroom design for aseptic environments
Ph. Eur. 5.1.6 Alternative methods for control of microbiological quality/ USP <1223> Validation of alternative microbiological methods	Rapid microbiological methods
USP <1115> Bioburden control of nonsterile drug products	Bioburden control
USP <1227> Validation of microbial recovery from pharmacopeial articles	Microbial method validation

In addition to the above, the chapters on water testing in the pharmacopeia include information relating to microbiological testing.

The microbiological test method and guideline general chapters (USP <61>, <62>, <71>, and <1111>, Ph. Eur. 2.6.1, 2.6.12, 2.6.13, and 5.1.4) are harmonized. That means that the basic text is the same for Ph. Eur, JP, and USP, and that tests conducted under one pharmacopeia are accepted by another.

Many of the above tests are culture based. Here microorganisms are grown in the laboratory by providing them with an environment suitable for their growth. The growth medium should contain all the correct nutrients and energy source and should be maintained at an appropriate pH, salinity, and oxygen tension and be free of antibacterial substances. Control of culture media, as detailed in Chapter 5, is of great importance.

Some of the different tests are discussed in more detail in the proceeding sections.

6.7   Microbiological examination of nonsterile products

6.7.1   Total viable aerobic count (Ph. Eur. 2.6.12, USP <61>)

These established tests are described in Chapter 8, for this reason they will not be outlined in detail here. In summary, the test is designed to count the number of microorganisms (as colony forming units, CFUs) in a nonsterile product or raw material. There are two parts: total aerobic microbial count (TAMC) and total yeast and mould counts (TYMCs).

With the method, a test sample is taken and processed (e.g., diluted and neutralized) and then:

- filtered and the filter placed on defined media (membrane filtration technique), or
- a sample aliquot is taken and placed in a Petri dish and specified media poured onto the sample (pour plate technique), or
- a sample aliquot is placed on the surface of defined media and smeared evenly over the surface (spread plate technique), or
- for mainly insoluble materials, sample dilutions are placed into a series of replicate tubes and the number of tubes showing growth give a statistical evaluation of the number of microorganisms in the sample (most probable number, MPN technique).

With these methods, it is necessary to demonstrate that the sample material, test reagents or any aspect of the test procedure, adversely affects the outcome of the test.

Before embarking on the test, it is important to compile information about the material to be tested such as: physical and chemical attributes, base formulation, and the estimated bioburden. This information helps considerably with experimental design, and test method selection and validation.

6.7.1.1   Bioburden determination

Tests for bioburden determination used, for example, for the examination of in-process material, are broadly similar to the TAMC method. The optimal counting range for colonies, on a 9-cm agar plate, is 20–250 CFU [5]. Consideration must also be given to incubation times and temperatures [6]. Further information on bioburden testing is outlined in Chapter 8.

6.7.1.2   Method validation

Bioburden tests, and the TAMC/TYMC tests, need to be qualified to show the appropriateness of the method to the material under test. This is particularly with the areas of sample preparation and the ability of the media to recover microorganisms in the presence of the test sample. The following four "validation" areas need to be considered:

(1) media growth promotion;
(2) sample preparation;
(3) test method;
(4) sample neutralization.

6.7.1.3   Media growth promotion

Media growth promotion is required to demonstrate that it supports growth and has the ability to detect organisms in the presence of the test sample [7]. This is achieved by using not more than 100 CFU of the following specified organisms.

(1) For tryptone soy agar (TSA) and tryptone soy broth (TSB):
   a. *Staphylococcus aureus* ATCC 6538 (NCIMB 9518, CIP 4.83, NBRC 13276);
   b. *Pseudomonas aeruginosa* ATCC 9027 (NCIMB 8626, CIP 82.118, NBRC 13275);
   c. *Bacillus subtilis* ATCC 6633 (NCIMB 8054, CIP 5262, NBRC 3134).
(2) For sabouraud dextrose agar (SDA):
   a. *Candida albicans* ATCC 10231 (NCPF 3179, IP 48.72, NBRC 1594);
   b. *Aspergillus niger* ATCC 16404 (IMI 149007, IP 1431.83, NBRC 9455).

Various equivalent strains can be used as obtained from approved culture collections. These are: American Type Culture Collection (ATCC), National Collection of Industrial and Marine Bacteria (NCIMB), Collection of Institute Pasteur (CIP), Imperial Mycological Institute (IMI), National Collection of Pathogenic Fungi (NCPF), and National Biologicals Resources Centre (NBRC).

For the media control, a comparison is made between the recovery of organisms from the specific media against the calculated inoculum. For the test method validation, a comparison is made between the recovery of organisms from media with test material with that of a diluent control. Results in both cases should not differ by more than 50%.

6.7.1.4   Sample preparation

This area for consideration and should prompt the questions:

(1) Where will the sample be prepared, such as on a laboratory bench, within unidirectional airflow cabinet or an isolator. They may also be safety concerns with require the use of a Microbiological Safety Cabinet (MSC). The primary purpose of an MSC is to protect the laboratory worker and the surrounding environment from pathogens. All exhaust air is high efficiency particulate air (HEPA) filtered as it exits the biosafety cabinet, removing harmful bacteria and viruses. This device is different from a unidirectional airflow cabinet, which blows unfiltered exhaust air toward the user and is not safe for work with pathogenic agents.
(2) In what will the sample be prepared? For example, a sterile plastic container.
(3) Agitation—does the sample need vortexing or sonication?
(4) Dilution—does the sample need diluting because of a high natural bioburden/because it is highly antimicrobial?

Some products are more complex to test and require pre-treatment. Examples of best practice for different products include:

Water-soluble products

• Dissolve or dilute (usually a 1 in 10 dilution is prepared) the product to be examined in phosphate buffer solution pH 7.2, If necessary, adjust to a pH of 6–8.
• Further dilutions, where necessary, are prepared with the same diluent to yield not more than 250 CFU/plate in case of TAMC, 50 CFU/plate in case of TYMC.

- For products or raw materials that do not dissolve completely, grind them in a sterile mortar and pestle, in an aseptic environment, to a fine powder.

Nonfatty products insoluble in water

- Suspend the product to be examined (usually a 1 in 10 dilution is prepared) in phosphate buffer solution pH 7.2.
- A surface-active agent such as 1 g/L of polysorbate 80 may be added to assist the suspension of poorly wettable substances.
- If necessary, adjust to a pH of 6–8. Further dilutions, where necessary, are prepared with the same diluent to yield not more than 250 CFU/plate in case of TAMC, 50 CFU/plate in case of TYMC.

Fatty products

- Dissolve in isopropyl myristate sterilized by filtration, or mix the product to be examined with the minimum necessary quantity of sterile polysorbate 80. Heated, if necessary, to not more than 40 °C or, in exceptional cases, to not more than 45 °C. Mix carefully and if necessary maintain the temperature in a water bath.
- Add a sufficient quantity of the pre-warmed chosen diluent to make a 1 in 10 dilution of the original product.
- Mix carefully, while maintaining the temperature for the shortest time necessary for the formation of an emulsion. Further serial tenfold dilutions may be prepared using the chosen diluent containing a suitable concentration of sterile polysorbate 80 or another noninhibitory sterile surface-active reagent.

6.7.1.5 Test method

Which test method is selected is based upon sample characteristics and the required microbial limits. The choice is between membrane filtration, pour plate, spread plate, and the MPN methods.

The limitations with these methods, and other culture-based assessments of bioburden, should be understood. The "colony count" is an indirect count with variable insensitivity, and it is very imprecise when very few colonies are counted. Furthermore, with the CFU, which is an artefact-based count relying on cellular replication to produce a visible speck of cells (the "colony") on the growth medium, if the medium or physical conditions are not adequate, then no colony appears. Another weakness is that if a clump of many cells lands in one place and only a single colony forms, then the count of "one" underestimates the total. Therefore, plate counts are not always precise or accurate [8]. The MPN, which is not a direct cell count, is also dependent on a cell's ability to multiply in growth medium under the physical incubation conditions.

To add to these issues, if the sample possesses antimicrobial activity that requires neutralization then there are principally three approaches to consider:

(1) chemical neutralization;
(2) enzymatic neutralization;
(3) dilution.

The appropriate method selected needs to be qualified.

Once the sample preparation has been developed and validated, the media performance qualified, and the test method selected then the routine test can be implemented.

6.7.2   Tests for specified organisms (Ph. Eur. 2.6.13, USP <62>)

The test for specific organisms is to determine the absence, or limited occurrence of specified microorganisms in a given pharmaceutical sample, that may be detected under the test conditions. These tests are described in Chapter 8, and they focus on screening for one or more of the following organisms:

(1) *Escherichia coli*: natural inhabitant of gut flora. Some species are pathogenic and cause diarrhoea. If recovered, the organism indicates fecal contamination;
(2) Salmonellae: common inhabitant of gut flora. If recovered, the organism indicates fecal contamination and of high pathogenicity;
(3) *S. aureus*: common inhabitant of human skin and nose, detectable in feces. If recovered it indicates high pathogenicity potential. There may also be a risk to product quality due to resistance to preservatives. The bacterium has a low nutrient demand can grow to high numbers in certain materials;
(4) *P. aeruginosa*: a common water inhabitant, especially of stored water. If recovered it indicates high pathogenicity potential;
(5) Clostridia: potential pathogens relating to specific situations, especially where anaerobic conditions are prevalent (e.g., with talc or bentonite);
(6) *C. albicans*: potential pathogens relating to specific situations, such as vaginal preparations.

Special, selective or differential agars are required for the examination of the above indicator microorganisms.

In addition to these described species, there may be a requirement to identify and test for specific "objectionable" microorganisms. These are undesirable organisms from a product quality/efficacy point of view or from a patient risk situation. Such organisms require the adoption of appropriate selective agars.

6.7.3   Specification limits (harmonized method)

The limits for total viable aerobic count and the tests for specified microorganisms are displayed in Tables 6.2 and 6.3.

It should be noted that the pharmacopeia allows variability in test results equal to a factor of two if the specified microbial limit is 10, the maximum accepted microbial count is 20 CFU and still meets the product specification, if the specified microbial limit is 100(10^2), the maximum accepted microbial count is 200 CFU and still meets the product specification and so on.

- 10^1 microorganisms: maximum acceptable count = 20.
- 10^2 microorganisms: maximum acceptable count = 200.
- 10^3 microorganisms: maximum acceptable count = 2000, and so forth.

Table 6.2 **Acceptance criteria for microbiological quality of nonsterile dosage forms and raw materials**

Route of administration	TAMC	TYMC	Specified microorganisms
Nonaqueous preparations for oral use	10^3	10^2	*Escherichia coli* absent in 1 g or 1 mL
Aqueous preparations for oral use	10^2	10^1	*Escherichia coli* absent in 1 g or 1 mL
Rectal use	10^3	10^2	*If required*
Oromucosal use Gingival use Cutaneous use Nasal use Auricular use	10^2	10^1	*Staphylococcus aureus* absent in 1 g or 1 mL *Pseudomonas aeruginosa* absent in 1 g or 1 mL
Vaginal use	10^2	10^1	*Staphylococcus aureus* absent in 1 g or 1 mL *Pseudomonas aeruginosa* absent in 1 g or 1 mL *Candida albicans* absent in 1 g or 1 mL
Inhalation use (special requirements apply to liquid preparations for nebulization)	10^2	10^1	*Staphylococcus aureus* absent in 1 g or 1 mL *Pseudomonas aeruginosa* absent in 1 g or 1 mL Bile-tolerant gram-negative bacteria absent in 1 g or 1 mL
Transdermal patches (limits for one patch including adhesive layers and backing)	10^2	10^1	*Staphylococcus aureus* absent in 1 g or 1 mL *Pseudomonas aeruginosa* absent in 1 g or 1 mL

Table 6.3 **Acceptance criteria for microbiological quality of raw materials for nonsterile manufacturing**

Material	TAMC	TYMC	Specified microorganisms
Substances for pharmaceutical use	10^3	10^2	*If required*

6.8 Measurement of cell concentration in suspension by optical density

A common issue for the microbiology laboratory is the determination of starting inoculum concentration. If the inoculum concentration is determined by plating, the inoculum is several days old before use, and if the inoculum is out-of-range, the resultant test will be invalid. A means to avoid this is to estimate the population through an assessment of cellular optical density.

To assess optical density, the most common method to use is spectrophotometry [9]. A spectrophotometer measures turbidity directly. The theory is that light passing through a suspension of microorganisms is scattered, and the amount of scatter is an indication of the biomass present in the suspension. Each spectrophotometer used must be independently calibrated for use in estimating microbial concentrations. The absorption of light is affected by the width of the instrument's slit, the condition of the filter, the size and condition of the detector, and the total output of the lamp [10].

With the use of optical density, the correlation of absorption to dry weight is very good for dilute suspensions of bacteria, and this relationship seems to hold regardless of cell size. In developing a method to estimate CFUs, a calibration curve is constructed. The calibration study must demonstrate the linear range of the absorbance against CFU values and the relevant values. It is important to note that, in more concentrated suspensions, this correlation (absorption to dry weight) does not correlate well.

6.9   Sterility testing

The sterility test applies primarily to finished products that are required to be sterile such as eye drops and intravenous products. The sterility test is a referee test; however, it is not intended as a sole release test. To verify sterility, other GMP expectations should be in place, controlled, and monitored such as environmental control, environmental monitoring, validated aseptic processing, validated sterilization processes, and so forth. The sterility test thus represents one set of data that contributes to the decision of whether or not the product lot meets the stated claims to be sterile. The test was first introduced in 1932 in the British Pharmacopeia as a direct inoculation test. This was followed by the USP in 1935. A membrane filtration version was introduced in 1957.

The test has several flaws. Most evidently, from Bryce's critique of the methodology. This highlighted the fact that "the sample size is so restricted that it provides only a gross estimate of the state of 'sterility' of the product lot" and that it "can only recognize organisms able to grow under the conditions of the test" [11]. Statistical evaluation of the sterility test indicates that it is limited use in assuring product sterility, only capable of detecting gross contamination. Moreover, the test will only detect those microorganisms that are capable of reproducing within the prescribed culture media and at the preselected temperature and for the described time period. Nevertheless, it is the recognized method, and it remains a component of any release strategy for a sterile product filled by aseptic processing (with terminally sterilized products, a case can be made for parametric release).

The sterility test is a qualitative, presence/absence test based upon growth for bacteria and fungi in two types of media. There are principally two methodologies applied which ask slightly different questions [12]:

1. Membrane filtration (funnel open method and, more commonly, a closed system method— the open funnel method poses a greater contamination risk).

With this method, the sample is filtered, then a rinse is undertaken to remove or to neutralize any product residues. Then, either two filters are used or a filter is cut filter in half. The two portions are placed into specified media and incubate at a defined temperature for 14 days.

The membrane filter requirements are:

(1) cominal pore size 0.45 μm;
(2) diameter about 50 mm;
(3) cellulose nitrate filters (e.g., for aqueous, oily, and weakly alcoholic solution);
(4) cellulose acetate filters (e.g., for strongly alcoholic solutions);
(5) other filter types may apply (e.g., antibiotics).

The question asked is: are there viable cells on or in the filter?

2. Direct inoculation

With this method, a sample is placed directly into specified media, containing neutralizers if required, and incubated at a defined temperature for 14 days.

The question asked is: are there viable cells in the sample?

The pharmacopeias indicate that, wherever possible, the membrane filtration technique is utilized due to the greater likelihood of recovery of contamination by virtue of the greater sample size. However, certain test articles, such as viscous oils, creams ointments, and medical devices, may not be filterable therefore direct inoculation will be the method of choice.

The following media is for sterility testing:

(1) Fluid thioglycollate medium: primarily for anaerobic bacteria, but will also isolate aerobic bacteria;
(2) Soya-bean casein digest medium (TSB): for the isolation of fungi and aerobic bacteria.

Other media can be used so long as they meet the requirements for growth promotion. Where neutralization is required, the neutralizer can be added to the media; for example, ß-lactamase (penase) for the testing of penicillins and cephalosporins. In the case of neutralizers, the type and volume required, and the efficacy/toxicity must be part of the sterility test validation. With penase, a specific validation using *S. aureus* is described.

All batches of media must be shown to promote growth. Specific organisms are described to demonstrate growth promotion:

a. Fluid thioglycollate medium:
  i. *Clostridium sporogenes, P. aeruginosa, S. aureus;*
b. Soya-bean casein digest medium:
  i. *A. niger, B. subtilis, C. albicans.*

Each medium is inoculated with not more than 100 CFU and incubated for 3 days (bacteria) or 5 days (fungi). Clearly visible growth must be observed.

Incubation conditions and times are as follows:

(1) fluid thioglycollate medium at 30–35 °C;
(2) soya-bean casein digest medium at 20–25 °C;
(3) for a total incubation period of 14 days with visual examination for turbidity.

With 14 days, Besajew demonstrated that 20% of all contaminants become visible between the 8th and 12th days after a retrospective evaluation of almost 8 years of data. Further, it was found that up to 10% of the time growth did not occur before the 11th and 12th days [13]. There were also issues around suboptimal growth conditions, inherent slow growers and injured cells.

There are two outcomes with the sterility test:

(1) clearly visible growth, which is equivalent in both the test sample and the control tubes indicates a valid sterility test process;

(2) if clearly visible comparable growth is not observed, then the developed sterility test method is not validated, and the test needs to be modified and the validation repeated.

With both the direct inoculation and membrane filtration methods, the sterility test is a demanding procedure where asepsis must be ensured to allow for correct interpretation of the results. Most importantly, the test environment must be adequate (like the production area). This requires the use of an EU GMP Grade A device with a Grade B background or a Grade A isolator operator in a Grade C or D cleanroom. To verify environmental acceptance, environmental monitoring should be undertaken during the monitoring session, and negative controls should be run during the test session.

The test samples should be representative of the batch of material under test, such as being drawn from the beginning, middle, and end of the aseptic fill process. The transfer of samples from the sampling area into the testing area and subsequent handling should be proceduralized. For example, the outside surfaces of vials should sanitized or gassed into the test area. The number of articles taken from a batch and the quantities required to be sterility tested are set out in the pharmacopeia. This relates to the quantity filled per container and to the batch size. For example, with parenteral preparations with batches of not more than 100 containers, the number of containers to be tested is 10% or four containers, whichever is the greater; whereas for more than 100 but not more than 500 containers, the 10 containers are tested; and with more than 500 containers then the number required is 2% or 20 containers, whichever is less *unless*, the product is a large-volume parenteral, in which case the number drawn from the batch is 2% or 10 containers, whichever is less.

6.9.1 Validating the sterility test

The test needs to be validated. For this, the characteristics of the material need to be considered, such as solubility and antimicrobial activity. This information is used to develop a validated sample preparation process. Dispersion of the material in diluents (where required) with or without surface active agents and neutralizers is essential for membrane filtration and direct inoculation to allow for adequate qualification of the sterility test.

Sterility test validation is multifactorial and requires:

(1) defined sample preparation;

(2) appropriately selected test method (i.e., membrane filtration/direct inoculation);

(3) media growth promotion and sterility studies;

(4) environmental control testing;

(5) operator validation;

(6) bacteriostasis/fungistasis effect of the product (now called the validation test in the harmonized methods).

The validation of the sterility test should be performed with the test articles using the developed sample preparation and selected methodology. Three different batches are normally tested. With the challenge microorganisms, these are the same as those used for media growth promotion. The challenge is fewer than 100 CFU. For membrane filtration, the specified organism is added to the final filter rinse and with direct inoculation, the specified organisms are added to the media. Growth must occur within 3 days for bacteria and within 5 days for fungi.

Rapid microbiological methods have been developed for sterility testing. These have yet to be adopted by the pharmacopeia, although the US Food and Drug Administration (FDA) accepts such methods as alternatives. Chapter 17 contains some information about rapid and alternative methods.

6.10   *In vitro* and *in vivo* testing for pyrogens and endotoxins

Pyrogens and endotoxins are a heterogeneous group of chemical entities that share the characteristic of (when injected) being able to cause fever. Pyrogens can be nonbacterial as well as bacterial in origin [14]. However, the main pyrogen encountered in the pharmaceutical industry is of Gram-negative bacterial origin. That is the lipopolysaccharide (LPS) from the bacterial cell wall. The test for bacterial endotoxins is described in detail in Chapter 11.

In terms of the range of pyrogenic substances, these are displayed in Table 6.4.

Tests for endotoxins are evaluated at various stages during pharmaceutical manufacturing such as water systems, raw materials, in-process steps, and finished product. With finished products, it is more commonplace to test finished products that are to be injected for endotoxin than it is to conduct a test for pyrogens (in the classic form of the rabbit pyrogen test). An alternative method is the monocyte activation test, which uses whole blood and involves the detection of cytokines.

6.10.1   Rabbit (in vivo) pyrogen test

The basis for the rabbit pyrogen test is that any pyrogen-containing solution injected intravenously will after a short period (circa 15 min) result in fever that peaks after about 90–120 min and then subsides [15]. The body temperature rise is proportional to the level of pyrogen. In reality, a measured dose of sample to be tested is injected into the ear veins of three rabbits. The cumulative rise in body temperature is then

Table 6.4 **Sources of pyrogens**

Nonbacterial	Bacterial
• Antigens (antibody mediated response) • Poly nucleotides • Steroids • Adjuvants (e.g., muramyl dipeptide) • Viruses • Fungi (yeast, polysaccharide capsules)	• Streptococcal toxins • Staphylococcal enterotoxins • Mycobacterial cell wall components • Bacterial cell wall: lipopolysaccharides (endotoxins)

periodically measured over a 3-h period via thermometers placed into the rectum of each rabbit. The summed temperature changes of the three rabbits is then compared against values representing "Pass," "Fail," and "Retest" acceptance criteria.

If a sample falls into the "Retest" criteria, then a further three rabbits may be tested, and this can be repeated up to four times (i.e., 12 rabbits). The rabbit test is considered to be about 50 times less sensitive that the *limulus* amoebocyte lysate (LAL) test. However, the rabbit sensitivity to pyrogens is similar to humans and, hence, does give an indication of the pyrogenic risk of the material to people. In addition, the rabbit test will react to all potential pyrogens not only LPS endotoxin.

6.10.2   LAL testing for bacterial endotoxin

It has long been known that the blood from the horseshoe crab (*Limulus polyphemus*) when in contact with Gram-negative bacteria becomes coagulated [16]. The mechanism for this coagulation occurs because in the presence of divalent cations (e.g., Ca^{2+}, Mg^{2+}) interaction with a "factor C" in the amoebocyte of the crab activates it. The active factor B induces a pro-clotting enzyme that converts the protein coagulogen into Coagulin resulting in coagulation. This mechanism is the basis of three LAL tests: gelation (or gel-clot) and two photometric methods: turbidimetric and chromogenic. These are outlined in Chapter 11.

6.11   Microbiological assay of antibiotics

The biological determination of antibiotic potency in pharmaceutical preparations is unchanged in principle since the 1950s. Antibiotic substances produced by fermentation are often controlled but representing a collection of closely related substances that individually may exhibit different biological activity [17].

The antibiotic bioassay provides a collective assessment of the potency of the overall biological activity of an antibiotic preparation. This activity (potency) is quoted in terms of international standards, specifically defined and quoted by pharmacopoeias. While many antibiotic assays have given way to chemical analysis such as high-performance liquid chromatography (HPLC), such methods do not reflect the true biological activity. Therefore, antibiotic bioassays still play an essential role in the manufacture and quality control of antibiotic medicines, but the assays still require a considerable amount of expertise and skill to ensure success.

6.12   Environmental monitoring

Microbiological environmental monitoring involves the collection of data relating to the numbers of microorganisms present in a clean room or clean zone. These microorganisms are recovered from surfaces, air, and people. Nonviable particle counting, a physical test, is often included within the program because this function has often resided with the microbiology department to perform and due to the theoretical relationship between high numbers of nonviable particles and viable counts.

The main aim of microbiological environmental monitoring is to assess the monitoring of trends over time and the detection of an upward or downward movement, within clean areas.

The viable count aspect of environmental monitoring consists enumerating the numbers of microorganisms present in a clean room by collection results by using the following sample types:

(a) passive air-sampling: settle plates;
(b) active air-sampling: volumetric air-sampler;
(c) surface samples: contact (RODAC) plates;
(d) surface samples: swabs;
(e) finger plates;
(f) plates of sleeves/gowns.

These methods, together with the environmental monitoring program, are described in detail in Chapter 16.

6.13 Water analysis

Microbiological water analysis is a method of analyzing water to estimate the numbers of bacteria present and to allow for the recovery of microorganisms in order to identify them.

The method of examination is the plate count. The plate count method relies on bacteria growing a colony on a nutrient medium, so that the colony becomes visible to the naked eye, and the number of colonies on a plate can be counted. Most laboratories use a method, whereby sample volumes of 100 mL (or greater) are vacuum filtered through purpose-made membrane filters, and these filters are themselves laid on nutrient medium within sealed plates [18]. A nonselective medium is used to obtain a total enumeration of the sample (called a heterotrophic plate count). When it is desirable to obtain a specific bacterial species, a selective medium can be used.

Sometimes testing requires an examination of indicator microorganisms. Indicator organisms are bacteria such as nonspecific coliforms, *E. coli* and *P. aeruginosa* that are very commonly found in the human or animal gut and which, if detected, may suggest the presence of sewage. Such organisms are detected using specialist agars or test kits.

Methods for water testing are described in Chapter 10.

6.14 Conclusion

This chapter has provided an overview of some of the common methods found within the microbiology laboratory. For those methods that are not discussed in detail elsewhere, the chapter has provided a general outline together with an indication of any methodological weaknesses. The general weakness pervading over all tests are with growing microorganisms and then with defining growth, both of which highlight the inherent variability that are commonplace to many microbiological techniques.

Despite the weakness, many of the methods are long established and can trace their methodological basis back to the experiments undertaken by the founding mothers and fathers of microbiology. The extent to which these methods will be replaced by rapid microbiological techniques is likely to be gradual, and, even then, it is unlikely the methods will disappear completely. Many will remain features of the microbiology laboratory for some time to come.

References

[1]   WHO. Good practices for pharmaceutical quality control laboratories. In: WHO Expert Committee on specifications for pharmaceutical preparations. Forty fourth report. Geneva: World Health Organization; 2010 [WHO Technical Report Series, No. 957. Annex 1].

[2]   Cross-Smiecinski AJ. Quality assurance. In: Hurst CJ, editor. Manual of environmental microbiology. Washington, DC: ASM; 2002. p. 158–65.

[3]   Arora DR. Quality assurance in microbiology. Indian J Med Microbiol 2004;22(2):81–6.

[4]   Cundell AM. Historical perspective on microbial method development. In: Easter MC, editor. Application and acceptance of rapid microbiological methods in the pharmaceutical industry. Boca Raton, FL: CRC/Interpharm Press; 2003. p. 9–17.

[5]   Zipkes MR, Gilchrist JE, Peeler JT. Comparison of yeast and mold counts by spiral, pour, and streak plate methods. J Assoc Off Anal Chem 1981;64:1465–9.

[6]   Sandle T, Skinner K, Yeandle E. Optimal conditions for the recovery of bioburden from pharmaceutical processes: a case study. Eur J Parenter Pharm Sci 2013;18(3):84–91.

[7]   Cundell A. Review of the media selection and incubation conditions for the compendial sterility and microbial limit tests. Pharm Forum 2002;28(6):2034–41.

[8]   Postgate JR. Viable counts and viability. In: Norris JR, Ribbons DW, editors. Methods in microbiology, vol. 1. New York: Academic Press; 1969. p. 611–28.

[9]   Koch AL. Growth measurement. In: Gerhardt P, editor. Methods for general and molecular bacteriology. Washington, DC: American Society for Microbiology; 1994. p. 248–77.

[10] Koch AL. Turbidity measurements of bacterial cultures in some available commercial instruments. Anal Biochem 1970;38:252–9.

[11] Bryce DM. Test for the sterility of pharmaceutical preparations. J Pharm Pharmacol 1956;8:561.

[12] Sandle T. Practical approaches to sterility testing. In: Saghee MR, Sandle T, Tidswell EC, editors. Microbiology and sterility assurance in pharmaceuticals and medical devices. New Delhi: Business Horizons; 2011. p. 173–92.

[13] Besajew VC. The importance of the incubation time in the test for sterility. Pharm Ind 1992;54(6):539–42.

[14] Hort E, Penfold WJ. A critical study of experimental fever. Proc R Soc Lond 1912;85:174–86.

[15] Seibert FB, Mendel LB. Temperature variations in rabbits. Am J Physiol 1923;67:83–9.

[16] Tours N, Sandle T. Comparison of dry-heat depyrogenation using three different types of Gram-negative bacterial endotoxin. Eur J Parenter Pharm Sci 2008;13(1):17–20.

[17] Sandle T. Introduction to antimicrobials. Pharmig News 2010;(Issue 39): 5–7.

[18] Sandle T, Skinner K. A practical example arising from the harmonization of the microbial enumeration method for water. Pharm Microbiol Forum Newsl 2009;14(4):2–5.

Bioburden determination

7.1   Introduction

This chapter is concerned with the examination of products (finished and interme-diate) and devices for bioburden. Bioburden is a term used to describe the microbial numbers on a surface (or complete item) or inside a device or from a portion of liquid. In the lexicon of microbiology, the term "bioburden" is somewhat misleading as "bur-den" implies that the level of microorganisms is automatically a problem or concern; whereas, in practice, the objective of qualitative or a quantitative testing is to ascertain if the types and numbers of microorganisms present are satisfactory or unsatisfactory when compared to predefined acceptance criteria [1].

In some literature, bioburden testing relates to raw materials testing, environ-mental monitoring, or in-process sample testing. These areas will have a given, even "natural" bioburden. When this bioburden rises above typical levels or ends up in the wrong place, then arguably the appropriate term to use is biocontamination [2]. Biocontamination, therefore, differentiates the population of microorganisms present as a problem, as distinct from the typical bioburden.

While bioburden can apply to the assessment of a number of microbiological attri-butes, in this chapter, bioburden will be limited to the assessment of nonsterile prod-ucts, in-process samples, and the assessment of product prior to sterilization. Thus, water testing and environmental monitoring are considered to be separate areas of microbiological monitoring, and they are described in separate chapters (although the reader should note that, in some literature, the term "bioburden" can embrace these types of testing).

In relation to the areas considered, this chapter looks at bioburden testing as a measure of the total number of viable microorganisms: that this, the total microbial count. Some reference is additionally made to the species of microorganism recovered (although as-sessing whether a species is potentially harmful is the subject of Chapter 8).

7.2   Total microbial count

The term "total microbial count" can refer to the total number of bacteria and fungi present or to the total number of bacteria. This confusion with the term has been enhanced by a pharmacopoeial chapter called the "microbial limits test" (which is harmonized between the European, US, and Japanese pharmacopeia). The chapter de-scribes a "total aerobic microbial count" (TAMC), which refers to bacteria only, and a "total yeast and mould count" (TYMC; which refers to fungi only). The two aspects of the microbial count—TAMC and TYMC—involve testing a sample on different agar and subjecting the tests to different incubation conditions [3]. In past editions of the pharmacopeia, the bacterial count and fungal count were added together to produce a total microbial count. However, since 2005, the pharmacopeia requires the two to be reported separately and assessed against specified limits.

Outside of the specifics of pharmacopeia testing, where bioburden is assessed as a quality control test to monitor a product manufacturing process (as with the case of in-process bioburden testing) invariably only one culture medium is used, and a total microbial count is assessed. This is conventionally referred to as either "total microbial count" or "total viable count." The latter term attempting to distinguish between live and dead microorganisms, or, more accurately those microorganisms that are capable of growing on the culture media used under the incubation conditions of the test. Viable is defined as the ability to multiply via binary fission under the controlled conditions. In contrast, in a microscopic evaluation, all cells, dead and living are counted.

Another alternative term is "total viable aerobic count" (or "total viable anaerobic count," depending upon the atmospheric conditions deployed during incubation). To an extent, these terms are interchangeable; what matters foremost is the method used to conduct the test. Knowing this makes the resultant data easier to interpret.

This discussion is not intended to be tautological, more to emphasize that there can be ambiguity in relation to the terms used to describe bioburden testing and clear definitions should be sought from the outset, especially when comparing test results between laboratories.

7.3   Units of measurement

With most methods of bioburden determination, the bioburden quantification is expressed in terms of colony forming units (CFUs). An exception to this is with the most probable number (MPN) method. Furthermore, Chapter 8 discusses the additional testing of samples for the presence or absence of specific microorganisms (this, when so required, and carried out according to the pharmacopeia is conducted as part of the microbial limits test). Here these microorganisms may or may not be enumerated for it may be sufficient, or the test itself is limited to, to note whether a particular microorganism of concern is present or not.

The CFU is an estimate of the number of viable bacteria or fungal cells in a sample. The visual appearance of a colony in a cell culture requires significant growth. With this it is unknown if the progenitor of the colony was one microorganism or several microorganisms. Hence, when counting colonies it is uncertain if the colony arose from one cell or 1000 cells, and importantly CFU is not a direct measure of microbial numbers. Results expressed in CFU are reported to a unit of measurement. Hence, results can be reported as colony-forming units per milliliter (CFU/mL) for liquids, or colony-forming units per gram (CFU/g) for solids to reflect this uncertainty (rather than cells/mL or cells/g). Therefore, an estimation of microbial numbers by CFU will, in most cases, undercount the number of living cells present in a sample [4].

Inaccuracies can also occur with the act of plate counting. Due to the size of the agar plate, there will be an optimal counting range, and errors will occur where microbial numbers are above an upper countable limit (due to confluence or overcrowding) or below a lower limit (due to statistical error in relation to accuracy of the count, particularly where dilutions have been performed) [5]. A further source of error can arise with rounding up or down or through averaging. For a comprehensive review of plate counting errors, Sutton has produced a seminal paper on the subject [6].

Another source of inaccuracy is due to many microorganisms in the environment being unculturable. By this, "unculturable" means, there are viable microorganisms present (and capable of renewed metabolic activity) but of a type that cannot grow on culture media (or at least the culture media and incubation conditions provided for by the test) [7]. The term can also apply to microorganisms that might, under a different set of conditions, grow on the medium, but they cannot because they have undergone physiological stress (such as starvation, elevated osmotic concentrations, exposure to white light and so forth), or they have been rendered sublethally damaged; conditions that prevent them from growing [8]. For these microorganisms, terms such as "viable but non-culturable" or "active but non-culturable" are deployed [9]. Numerous bacteria, both Gram-positive and Gram-negative, pathogenic and nonpathogenic, can enter such a state [10].

The count obtained is also dependent upon the extent that culturable microorganisms can grow under the test conditions. This is a factor of the type of culture media used, the atmospheric conditions, the temperature of incubation, and the time that the cultivated medium is incubated for. Importantly, no testing scheme can detect everything. Thus, tests and test results have the objective of providing the best indicator possible of the microbial bioburden *but not* the absolute bioburden. In many cases, test regimes are biased toward aerobic, mesophilic microorganisms. This is because such organisms are common to the environment; they will often be a problem should they contaminate the product since they are the most likely to grow; and because most human pathogens fall within this grouping [11].

Inaccuracies and error can occur with colony counting. The counting of colonies manually is normally carried out using an artificial light source, such as a colony counter. As an alternative, several automated colony counters are available. Automated systems can be difficult to validate and can experience problems when attempting to differentiate plates with colonies from a range of different microbial species [12].

7.4   Nonsterile products and microbial limits testing

The microbiological quality of the finished product is determined by the quality of the starting materials; materials with a known low bioburden should be purchased whenever possible. These are, in most cases, "nonsterile" materials or products. Such materials are either used to manufacture more complex pharmaceutical products (such as tablets, creams, or ointments) or used in the preliminary stages of what will become sterile products.

Nonsterile products are assessed, according to the harmonized pharmacopeia, for total count (a separate bacterial and fungal count is required). In addition, for some products, additional testing is required for indicator microorganisms. This is covered in Chapter 8 (the chapter extends the discussion of prescribed indicator microorganisms to any additional ones that are considered to be "objectionable" based on a risk assessment).

With the pharmacopeia, if a material monograph requires a test for microbial limits, then Ph. Eur 2.6.12 or USP <61> is applied (what is termed the microbial limits test) [13]. With such products or constituent raw materials that are used for the manufacture of sterile products, the microbial limit for these materials must not exceed 10^2–10^3 microorganisms per milliliter (as set out in the harmonized pharmacopeia, with the USP this is chapter <1111> and with the European pharmacopeia this is Chapter 5.1.4). Raw materials are defined as those substances that can be brought into a manufacturing unit either for further processing or to aid in such processing. In the microbiological control of pharmaceutical raw materials, there is one primary aim—to exclude any microorganism that may subsequently result in deterioration of the product or may harm the patient.

The compendial microbial limits test is made up of two parts [14]:

- TAMC. This is an estimation of viable aerobic mesophilic microorganisms that can be derived from a general purpose medium (the pharmacopeia recommends soya bean casein digest medium).
- TYMC. This is an estimation of mesophilic aerobic fungi (yeast like and filamentous, and those that are dimorphic). The test uses a general purpose fungal medium (the pharmacopeia recommends Saboraud dextrose agar).

The bioburden test is either one or both of the compendial TAMC or TYMC methods, or an alternative. For the examination of the microbial count, there are four recommended methods. These methods are outlined below. The methods described are the variants according to the pharmacopeia, which are required to claim that the microbial limits test has been conducted. Where the pharmacopeia is not required to be followed (such as for in-process bioburden testing, as discussed later), then variations to these methods can be conducted if appropriately justified.

7.4.1   Membrane filtration

This is the method of choice and should be applied to samples that contain antimicrobial substances. With the method, the sample is passed through a membrane filter with a pore size of 0.45 μm or less. Filters about 50 mm across are recommended, but other sizes may be used. Usually, the test measures two test fluids of 10 mL each, passing each sample through a separate filter. It is important to dilute the pretreated test fluid if the bacteria concentration is high, so that 10–100 colonies can develop per filter. After filtration, each filter must be washed three times or more with an appropriate liquid such as phosphate buffer, sodium chloride-peptone buffer or fluid medium. The volume of the washings should be about 100 mL each. If the sample includes lipid, polysorbate 80 or an appropriate emulsifier may be added to the washings.

After filtration, for bacteria detection, the two filters must be placed on a plate of soybean–casein digest agar medium, and for fungi detection, an antibiotic is added to the medium and placed onto a plate of one of the Sabouraud glucose agar. Plates are incubated for at least for 5 days at 30–35 °C for bacteria detection and at 20–25 °C for fungi detection. At the end of the incubation period, the number of colonies is counted.

7.4.2 Direct plating methods

There are two direct plating methods: pour plate and spread plate. Of the two, the pour plate is preferential because of a greater theoretical accuracy.

7.4.2.1 Pour plate method

With the pour plate method, Petri dishes of 9–10 cm in diameter are used, with two agar media used for each dilution. For the test:

- Take 1 mL of the test fluid or its dilution into each Petri dish aseptically, add to each dish 15–20 mL of sterilized agar medium, previously melted and kept below 45 °C, and mix (45 °C is just above the point of solidification to minimize heat-induced cell death).
- For bacteria detection, use soybean–casein digest agar medium and for fungi detection, use Sabouraud glucose agar media, to which antibiotic has previously been added.
- After the agar solidifies, incubate at least for 5 days at 30–35 °C for bacteria detection and at 20–25 °C for fungi detection. If a large number of colonies develop, calculate viable counts based on counts obtained from plates with not more than 300 colonies per plate for bacteria detection and from plates with not more than 100 colonies per plate for fungi detection.

7.4.2.2 Spread plate method

With the spread plate method:

- Place 0.05–0.2 mL of the test fluid on the solidified and dried surface of the agar medium and spread it uniformly using a spreader.
- Proceed under the same conditions as for the pour plate method, especially with regard to Petri dishes, agar media, incubation temperature and time, and calculation method.
- A variant, not described in the pharmacopeia, is the drop-plate method (or Miles and Misra method), wherein a very small aliquot (usually about 10 μL) of sample from each dilution in series is dropped onto a Petri dish [15].

7.4.3 Most probable number method

The MPN method (alternatively, the method of Poisson zeroes) is a method of obtaining quantitative data on concentrations of discrete items from positive/negative (incidence) data. The method involves taking the original solution or sample, and subdividing it by orders of magnitude (frequently 10× or 2×) into culture broth, and assessing the presence/absence in multiple subdivisions. The major weakness of MPN methods is the need for large numbers of replicates at the appropriate dilution to narrow the confidence intervals [16]. The MPN is only effective of the examination of bacteria, as it does not provide reliable results for the enumeration of fungi.

To determine the accuracy and sensitivity of the test methods used for microbial limit testing, 10 g or 10 mL samples of the test material are examined. It is also important, when conducting these tests, to ensure that the testing method does not either introduce bacteria into the test sample or kill bacteria in the test sample [17]. Furthermore, the dilution of microbial challenges needs to be as precise as possible [18].

7.4.4   Method verification

Method verification is an important step. While the total count method is a "compendia test," which means that, by convention, the test itself does not require validating, the suitability of each material must be qualified to show that the test method is not inhibitory and that any microorganisms present can be recovered. This assessment is particularly important for samples that have antimicrobial activity.

When test samples have antimicrobial activity or when they include antimicrobial substances, these antimicrobial properties must be eliminated by dilution, filtration, neutralization, inactivation, or other appropriate means. The tests should be conducted for samples prepared by mixing multiple portions randomly chosen from individual ingredients or products. When samples are diluted with fluid medium, the tests must be conducted quickly. Due attention must be paid to the effective quality control and the prevention of biohazard.

The compendial chapters outline useful information for the development process as well:

* if the product contains antimicrobial activity, this should be neutralized;
* if inactivators are used, their efficacy and their absence of toxicity for microorganisms must be demonstrated;
* common neutralizing agents and methods include the addition of polysorbate or lecithin, and/or dilution methods.

7.5   In-process material bioburden assessment

The environmental and process bioburden should be monitored to ensure that they are both within acceptable limits. Environmental monitoring is addressed in Chapter 16; the focus here is the bioburden monitoring of intermediate product as the product moves through the manufacturing stages ("in-process" monitoring). The requirement for this step is outlined in the Code of Federal Regulations. 21 C.F.R. 211.110 (a)(6), which states that bioburden in-process testing must be conducted pursuant to written procedures during the manufacturing process of drug products.

EU good manufacturing practice (GMP) does not specifically address in-process sampling. Nonetheless, selecting and examining samples for bioburden determination using a total viable count method is commonplace. With bioburden testing, an appropriate test method should be selected. Here either membrane filtration (using a 100 mL sample) or the pour plate (using a 1 mL sample) are the most common methods. The pharmacopoeial methods, as outlined above, do not need to be followed (e.g., it is more common to use one general purpose agar, such as soybean–casein digest medium, rather than two different agars and to incubate samples between 20 and 35 °C). Whichever method is adopted, it is important that the method is assessed as suitable. Here, as with the microbial limits test, an assessment must be made of the presence of any antimicrobial substances.

In establishing an in-process bioburden regime, appropriate limits should be set (in the form of "alert" and "action" levels). The levels can be defined as [19]:

- Alert level: a level that, when exceeded, indicates a process may have drifted from its normal operating condition. Alert levels constitute a warning, but do not necessarily warrant corrective action.
- Action level: a level that, when exceeded, indicates a process has drifted from its normal operating range. A response to such an excursion should involve a documented investigation and corrective action.

These limits should decrease as the process moves downstream. While limits will relate to specific processes, a general guideline is 100 CFU for the start of the process and 10 CFU near the end of the process (per 1 mL of per 100 mL). With sterile products, where a liquid is subject to a terminal sterilizing filter, within Europe there is a requirement for the liquid to contain no more than 10 CFU/100 mL.

When setting "alert" and "action" limits, it is good practice to:

- base levels on historical data;
- use means and/or standard deviations;
- perform continued trend analysis and data evaluation to determine if the established levels remain appropriate;
- watch for periodic spikes, even if averages stay within levels.

In addition to the monitoring of in-process samples, steps should be taken to ensure that control is maintained. Where known risks are apparent, specific process steps should be included to reduce these potential risks to a level consistent with the level of control required. Keeping control means that alert and action level excursions should be investigated. At the alert level, the following items could be considered, although generally no action or investigation is required:

- numbers and types of routine bioburden trends (product and environment);
- identification of recovered microorganisms;
- evaluation of microorganism for resistance to the sterilization process;
- production personnel impact (e.g., proper training or new personnel);
- manufacturing process changes;
- sampling and testing procedures changes;
- evaluation of laboratory controls and monitors;
- additional testing;
- thorough cleaning of production area;
- modification of sampling plan;
- raw materials and supplier changes;
- water-source contamination.

At the action level, the above-mentioned items are usually required in addition to the following:

- root cause analysis/investigation;
- determination of potential impact on sterilization specifications.

7.6   Presterilization bioburden assessment

For sterile products, bioburden assessment is a key requisite prior to sterilization. This is necessary of terminally sterilized products and products that are to be aseptically filled.

7.6.1   Terminally sterilized products

With terminally sterilized products, understanding the bioburden is necessary because the extent of the treatment of a sterilization process is a factor of the typical bioburden on or in the product; the resistance of the microorganisms that make up the bioburden; and the sterility assurance level required [20]. The test is important because an underestimation of the bioburden population could result in a miscalculation of the sterilizing requirements for a given product; in contrast, an overestimation could result in excessive exposure to the sterilizing agent, which in turn could affect the quality of the product.

At the stage prior to terminal sterilization, bioburden can be regarded as the sum of the microbial contributions from a number of sources including raw materials, manufacturing of components, assembly processes, manufacturing environment, assembly/manufacturing aids (such as compressed gases, water, and lubricants), cleaning processes, and packaging of finished product.

All of this matter because during the assessment of the suitability of the sterilization process, each product will have been assessed against a particular sterilization method (such as dry heat, most heat, radiation, or gas) for a particular dose (such as temperature or radiation dose) and for a given time. This assessment will have been made either by testing various representative items of the product for bioburden or using biological indicators of a known population, species and resistance. If the bioburden on or in a given product exceeds the bioburden assessed during the initial qualification then, theoretically, some microorganisms might survive.

Whether assessing the bioburden alone is a sufficient control measure is something the microbiologist must decide. If by designing the manufacturing process, the presterilization bioburden is kept consistently and reliably at very low numbers, a direct bioburden-based cycle may be a possible approach. Bioburden-based cycles are where the bioburden is assessed prior to each individual sterilization cycle being run. This is as a replacement to the use of biological indicators [21].

7.6.2   Aseptically filled products

With aseptic processing, one of the most important samples is taken from the bulk material prior to transfer through a sterilizing grade filter in preparation for aseptic filling. The filters used are generally of a pore size of 0.22 µm, configured in a series [22]. The "sterilizing filter" was defined in 1987 by the US Food and Drug Administration (FDA) on the basis of its retaining a minimum of 1×10^7 CFUs of *Brevundimonas diminuta* per square centimeter of effective filtration area (EFA).

Within Europe, there is a requirement for the challenge liquid to contain no more than 10 CFU/100 mL. In other territories, the permitted number of microorganisms is determined by risk assessment.

While bioburden assessment is important, aseptic processing carries continued risks. The sterile filtered liquid must subsequently be dispensed into sterile containers under a protective airflow. At this stage, contamination can be introduced if controls are not properly maintained.

7.6.3 Medical devices

A third area for bioburden determination prior to sterilization is in relation to medical devices. As with terminally sterilized pharmaceutical products, the objective is to ensure that the presterilization bioburden is below that used to qualify the sterilization process. Bioburden testing for medical devices made or used in the United States is governed by Title 21 of the Code of Federal Regulations and worldwide by the standard ISO 11737.

Bioburden control with medical devices, as described in ISO 11737, consists of five key steps. These are:

1. sample selection;
2. the process of removing any microorganisms from the sample;
3. transfer of microorganisms to recovery solutions;
4. enumeration of microorganisms;
5. data interpretation (with the application of correction factors, where necessary).

7.7 Alternative methods of bioburden assessment

As with many established fields of microbiology, rapid and alternative microbiological methods are available. Alternative methods include the use of polymerase chain reaction (PCR) assays, where deoxyribonucleic acid (DNA) can be extracted and examined using specific primers. The advantages conferred by PCR are specificity toward target microorganisms and faster time-to-result [23]. A different approach is with fluorescent-based technology, where cells are labeled and fluorescence detected through imagers. The reaction requires active microbial metabolism for enzymatic cleavage of a nonfluorescent substrate. Once cleaved inside the cell, the substrate liberates free fluorochrome into the microorganism cytoplasm. As fluorochrome accumulates inside the cells, the signal is naturally amplified. The cells are then exposed to the excitation wavelength of the fluorescent dye in a reader so that they can be visually counted. A variation upon this is with the use of digital imaging technology that automatically enumerates microcolonies earlier than the traditional visual plate counting methods allow. Such systems capture the native fluorescence (autofluorescence) that is emitted by all living cells.

7.8   Conclusion

This chapter has considered the subject of bioburden (the degree of microbial contamination or microbial load; the number of microorganisms contaminating an object) and bioburden testing. As the chapter has discussed, "bioburden" is not a straightforward concept. Complications arise in relation to the recovery of microorganisms and the fact that the "CFU" is a mere estimation of the numbers of microorganisms present.

Bioburden testing is applied to materials and products as per the internationally harmonized pharmacopeia under the guidelines pertaining to the microbial limits test. Here methods and agars, along with incubation conditions, are precisely defined. Beyond the pharmacopoeial test, bioburden assessments are required for intermediate product (in-process material) and, with sterile products, at the point closest to sterilization. Thus, bioburden estimation stands as an important and necessary part of pharmaceutical microbiology.

References

[1]   Hodges N. Bioburden determination. In: Halls N, editor. Microbiology and contamination control in pharmaceutical clean rooms. Boca Raton, FL: CRC Press; 2004. p. 115–36.

[2]   Cundell AM. Managing the microbiological quality of pharmaceutical excipients. PDA J Pharm Sci Technol 2005;59(6):381–95.

[3]   Bryans T, Alexander K. A comparison of incubation periods for bioburden isolates. Biomed Instrum Technol 2006;40:67–71.

[4]   Weenk GH. Microbiological assessment of culture media: comparison and statistical evaluation of methods. Int J Food Microbiol 1992;17:159–81.

[5]   Clarke JU. Evaluation of censored data methods to allow statistical comparisons among very small samples with below detection limit observations. Environ Sci Technol 1998;32:177–83.

[6]   Sutton S. Accuracy of plate counts. J Validation Technol 2011;17(3):42–6.

[7]   Bogosian G, Bourneuf EV. A matter of bacterial life and death. EMBO Rep 2001;2(9):770–4.

[8]   Oliver JD. The viable but nonculturable state and cellular resuscitation. In: Bell CR, Brylinsky M, Johnson-Green P, editors. Microbial biosystems: new frontiers. Halifax, NS: Atlantic Canada Society for Microbial Ecology; 2000. p. 723–30.

[9]   Xu H-S, Roberts N, Singleton FL, Attwell RW, Grimes DJ, Colwell RR. Survival and viability of nonculturable *Escherichia coli* and *Vibrio cholerae* in the estuarine and marine environment. Microb Ecol 1982;8:313–23.

[10] Kell DB, Kapreylants AS, Weichart DH, Harwood CL, Barer MR. Viability and activity in readily culturable bacteria: a review and discussion of the practical issues. Antonie Van Leeuwenhoek 1998;73:169–87.

[11] Nyström T. Not quite dead enough: on bacterial life, culturability, senescence, and death. Arch Microbiol 2001;176(3):159–64.

[12] Clarke ML, Burton RL, Hill AN, Litorja M, Nahm MH, Hwang J. Low-cost, high-throughput, automated counting of bacterial colonies. Cytometry A 2010;77(8):790–7.

[13] Clontz L. Microbial limit and bioburden tests. In: Clontz L, editor. Microbial contamination and control. Buffalo Grove, IL: Interpharm Press; 1998. p. 11.

[14] Clontz L. Microbial limit and bioburden tests: validation approaches and global requirements. Boca Raton, FL: CRC Press; 2009, p. 66–7.

[15] Miles AA, Misra SS, Irwin JO. The estimation of the bactericidal power of the blood. J Hyg 1938;38(6):732–49.

[16] Oblinger JL, Koburger JA. Understanding and teaching the most probable number technique. J Milk Food Technol 1975;38(9):540–5.

[17] Hoxey E. Validation of methods for bioburden estimation. In: Morrissey RF, editor. Sterilization of medical products. Morin Heights, QC: PolyScience; 1993. p. 22–42.

[18] Hedges AJ. Estimating the precision of serial dilutions and viable bacterial counts. Int J Food Microbiol 2002;76(3):207–14.

[19] PDA. Fundamentals of an environmental monitoring program. PDA technical report no. 13, vol. 55(No. 5); September/October 2001.

[20] Booth AF. Industrial sterilization technologies: principles and overview. In: Halls N, editor. Pharmaceutical contamination control. Bethesda, MD: PDA/DHI; 2007. p. 195–220.

[21] Russell AD. Destruction of bacterial spores by thermal methods. In: Russell AD, Hugo WB, Ayliffe GAJ, editors. Principles and practice of disinfection, preservation and sterilization. 3rd ed. Oxford: Blackwell Scientific Publications; 1999. p. 640–56.

[22] Caldwell M, Helt B, Holden B, McBride F, Schreier K. Aseptic manufacturing facility design. In: 3rd ed. In: Nema S, Ludwig JD, editors. Pharmaceutical dosage forms: parenteral medications, vol. 2. London: Informa Healthcare; 2010. p. 1–55.

[23] Jimenez L, Small S, Ignor R. Use of PCR analysis for detecting low levels of bacteria and mold contamination in pharmaceutical samples. J Microbiol Methods 2000;41(3):259–65.

Specified and objectionable microorganisms

8.1 Introduction

This chapter focuses on the presence of specific microorganisms in active pharmaceutical ingredients, pharmaceutical products, or raw materials that might, under certain conditions, be classed as "objectionable." In many ways, the chapter is a companion chapter to Chapter 7 on bioburden determination. With specific and objectionable microorganisms, the screening and examination for such organisms normally goes hand-in-hand with bioburden testing. This is because tests for nonsterile pharmaceutical products and ingredients (such as raw materials) involve assessment of total counts and presence/absence of particular organisms of concern.

Specific microorganisms (specified microorganisms) are described in the internationally harmonized pharmacopeia (Ph. Eur. 2.6.13 and USP <62>; which harmonized in 2006) [1]. These organisms and their significance are discussed in the chapter; however, the chapter does not seek to simply regurgitate the pharmacopeia, and the reader is referred to current compendia for the test method. With the case of objectionable microorganisms, although the pharmacopeia define certain "index" or "indicator" microorganisms, contemporary approaches to risk assessment require the microbiologist to define a wider list of organisms of concern (indeed USP <1111> drew this concerns to attention in 2006) [2]. This list cannot be defined as a general selection of microorganisms for inclusion relates to specific types of products and the intended patient population for those products [3].

Thus, the concept of objectionable microorganism consists of an array that are divided between specific indicator microorganisms required for qualitative testing by the pharmacopeia and those defined as objectionable by the pharmaceutical organization in relation to a particular product.

8.2 Indicator microorganisms

Chapter 7 describes the examination of nonsterile pharmaceutical products and constituent ingredients microbial numbers (bioburden). For certain materials, there is a compendial requirement for the absence of certain microorganisms [4]. These specified microorganisms include pathogens, such as *Salmonella*, and indicators of fecal contamination, such as *Escherichia coli*. These microorganisms are specifically listed because they directly, or they may indicate the presence of other microorganism from similar sources that pose a particular risk to immunocompromised patients [5]. This is because small numbers of opportunistic pathogens become infectious when the

Pharmaceutical Microbiology. http://dx.doi.org/10.1016/B978-0-08-100022-9.00008-6

body's resistance mechanisms become impaired, through disease or as a consequence of courses of immunosuppressant drugs [6]. Indeed the risk is such that, as modeling has demonstrated, it is impossible to rule out the possibility that single pathogenic microorganism, when ingested, has the potential of inducing infection and disease [7].

These specified microorganisms are intended to be indicators of wider contamination of a type that poses a risk to human health. In essence, this means that although particular microorganisms are *specified*, there could be other microorganisms of concern that may be found in similar niches to those listed. Therefore, while it could be possible to risk assess the mere presence of a specified microorganism should it only be recovered from the sample in low numbers, the mere presence of the organism could be indicative of other microorganisms of concern to human health.

The full list of specified microorganisms described in the harmonized pharmacopeia (USP <62> and Ph. Eur. 2.6.13) is:

- bile-tolerant Gram-negative bacteria;
- *E. coli*;
- *Salmonella*;
- *Pseudomonas aeruginosa*;
- *Staphylococcus aureus*;
- *Clostridia*;
- *Candida albicans*.

These species or types of microorganisms represent:

(a) Bile-tolerant Gram-negative bacteria

With this category, the pharmacopeia have chosen a diverse grouping, and one ill-defined since there is no strict definition of this group of microorganisms. Bile-tolerant Gram-negative bacteria are best defined as those microorganisms that show growth in the stated conditions on violet red bile glucose agar medium (thus the definition is, somewhat anachronistically, centered on a culture medium). They include those Gram-negative bacteria that grow in the presence of bile salts, which are nonlactose fermenting but at the same time able to utilize glucose. Examples of some bile tolerant Gram-negative bacteria includes members of the Enterobacteriaceae and of the genus *Pseudomonads* and *Aeromonas*. In keeping with imprecise definition, there is no clear consensus as to what defines "Enterobacteriaceae" [8]. The old-fashioned categorization was of "enteric bacteria," and later of gammaproteobacteria. Conventionally this grouping includes pathogens, such as *Salmonella*, *E. coli*, *Yersinia pestis*, *Klebsiella*, and *Shigella*. Other disease-causing bacteria in this family include *Proteus*, *Enterobacter*, *Serratia*, and *Citrobacter*.

With the pharmacopeia described test, not less than 1 g of the product is enriched with an Enterobacteria enrichment broth mossel, and after incubation at 30–35 °C for a defined time, a sub-culture is performed onto violet red bile glucose agar medium.

(b) *E. coli*

E. coli is a Gram-negative, facultatively anaerobic, rod-shaped bacterium of the genus *Escherichia* that is commonly found in the lower intestine of warm-blooded organisms (endotherms). Most *E. coli* strains are harmless, but some serotypes can cause serious food poisoning in their hosts and are occasionally responsible for product recalls due to food contamination (and on very rare occasions, pharmaceutical products) [9].

With the pharmacopeia test, both MacConkey broth and MacConkey agar are used to examine for the presence of *E. coli*.

(c) *Salmonella*

Salmonella is a genus of rod-shaped, Gram-negative bacteria. There are only two species of *Salmonella*, *Salmonella bongori*, and *Salmonella enterica*, of which there are around six subspecies and innumerable serovars. They can be divided into two groups—typhoidal and nontyphoidal Salmonella serovars. Nontyphoidal serovars are more common and usually cause self-limiting gastrointestinal disease. Typhoidal serovars include *Salmonella typhi* and *Salmonella* Paratyphi A, which are adapted to humans and do not occur in other animals [10].

The pharmacopeia test for *Salmonella* involves the use of Rappaport Vassiliadis Salmonella enrichment broth and xylose lysine deoxycholate agar.

(d) *P. aeruginosa*

P. aeruginosa is a Gram-negative, aerobic, coccobacillus bacterium with unipolar motility. It is an opportunistic human pathogen, often associated with contaminated water systems. *P. aeruginosa* typically infects the pulmonary tract, urinary tract, burns, wounds, and causes blood infections. The organism is fairly straightforward to identify for *P. aeruginosa* that secretes a variety of pigments, including pyocyanin (blue-green), pyoverdine (yellow-green and fluorescent), and pyorubin (red-brown) [11]. According to the pharmacopeia, the recommended agar for isolation and differentiation is cetrimide agar.

(e) *S. aureus*

S. aureus is a Gram-positive coccal bacterium that is frequently found in the human respiratory tract and on the skin. While *S. aureus* is not always pathogenic, it is a common cause of skin infections (e.g., boils), respiratory disease (e.g., sinusitis), and food poisoning. Disease-associated strains often promote infections by producing potent protein toxins [12]. With the pharmacopeia, the agar used for detection is mannitol salt agar.

(f) Clostridia

The Clostridia are a class of Firmicutes, including *Clostridium* and other similar genera. They are distinct from the genus *Bacillus* through lacking aerobic respiration. *Clostridium* are rod-shaped, Gram-positive endospore-forming bacteria. There are a number of species that can cause disease in humans, including *Clostridium botulinum*, *Clostridium difficile*, and *Clostridium tetani*. All pathogenic clostridial species produce protein exotoxins (such as botulinum and tetanus toxins) that play an important role in pathogenesis [13].

The pharmacopoeial test method deploys reinforced medium for Clostridia followed by Columbia agar.

(g) *C. albicans*

C. albicans is a diploid fungus that grows both as yeast and filamentous cells. It is a causal agent of opportunistic oral and genital infections in humans. *C. albicans* is commensal and a constituent of the normal gut flora comprising microorganisms that live in the human mouth and gastrointestinal tract. The fungus becomes a risk in the immunocompromised host [14].

With the pharmacopoeial method, both Sabouraud dextrose broth and Sabouraud dextrose agar are used for isolation.

8.2.1 Pharmacopeia methods

Under each section, the pharmacopeia states how much of the product or excipient is to be examined and how to incubate with each type of media with product in order to isolate any of the potential "specified" microorganisms within that product. These specified microorganism challenges must be validated to recover microbial growth as well. This portion of the microbial limits test is a presence/absence test. Depending on the product or excipient, one may choose to validate any number of the specified microorganisms from the pharmacopeia.

These bacterial and fungal indicators are selected as representatives of microorganisms that may cause disease in immunocompromised people or in other classes of susceptible persons. If such microorganisms were present, whether infection occurs, and the form it takes, depends on the route of administration, the dose of organisms, and the class of person.

Not all of these microorganisms require testing for; those that are required are described in individual monographs. This is in recognition that some types of nonsterile products are more prone to contamination than others. This reflects the point of origin or method of manufacture of the products. For example, one product prone to contamination is Arabic gum [15].

The pharmacopeia requires that where one or more of these microorganisms is to be examined, this is by qualitative analysis (a "presence–absence" test). For this, a portion of the sample (10 g or 10 mL) is incubated in broth for at least 24 h in order to enhance the isolation of any microorganisms present. The reason for incubating the samples for at least 24 h is due to the organisms, if they are present, being so in lower numbers than other types of microorganisms (for this reason identifying what is recovered for a bioburden test is insufficient since the microorganisms of concern may have simply failed to grow). An enrichment step and growth on selective media will enhance the isolation of pathogenic microorganisms.

8.2.2 Method qualification

Before sample testing is performed, the methods must be shown to be capable of detecting and isolating the specified microorganism of concern. This part of the procedure is called the preparatory testing. The preparatory testing involves the inoculation of different types of microorganisms into the samples to demonstrate the accuracy, efficacy, reproducibility, and sensitivity of a given method for detecting microbial contamination. With some products, a pretreatment test may be necessary depending upon the physical state of the product. Semisolid materials, for instance, need to be treated in order to form a solution or suspension.

It must also be established that the culture media for the test is suitable. This is affirmed by challenging each medium with a suitable panel. A test panel will include those microorganisms that should grow on the medium; microorganisms where growth on the medium reveals particular indicative properties, such as certain colonial pigmentation; and microorganisms that should not be recovered on the medium because the medium is intended to be inhibitory. For example, taking mannitol salt agar, which is used for the test for *S. aureus*, then the appropriate control organism for growth promotion and indicative growth is, unsurprisingly, *S. aureus*; whereas the organism for the test for inhibition is *E. coli* (where *E. coli* should not be recovered on the agar).

8.3   Determining which microorganisms are objectionable and assessing risk

Rigidly testing for the microorganisms listed in the compendia may not be the correct strategy. This is because it is recognized that there may be other "objectionable" microorganisms that are more appropriate and pose a greater risk to the product and therefore to the patient [16]. With this regard, the harmonized pharmacopeia requires that the significance of other microorganisms recovered should be evaluated. This is also in keeping with the requirements of the FDA Code of Federal Regulations (CFR). These missives are:

- 21 CFR 211.84(d)(6)—"Each lot of a component, drug product container, or closure with potential for microbiological contamination that is objectionable in view of its intended use shall be subjected to microbiological tests before use."
- 21 CFR 211.113(a)—"Appropriate written procedures, designed to prevent objectionable microorganisms in drug products not required to be sterile, shall be established and followed."
- 21 CFR 211.165(b)—"There shall be appropriate laboratory testing, as necessary, of each batch of drug product required to be free of objectionable microorganisms."

In keeping with risk assessment methodologies, it is incumbent upon the pharmaceutical organization to define their own problematic microorganisms [17]. Assessing whether a microorganism is objectionable requires the assessment of a number of factors. The foremost factor is whether the organism is a pathogen for if the organism is known to be a pathogen, and the route of infection is the same as the route of administration for the product, the organism is most likely objectionable.

Evaluation of whether a microorganism is or is not objectionable should include the following:

- The use of the product and the method of application (eye, nose, respiratory, dermal, and so on) in relation to different microorganisms. This is because different microorganisms carry differing risks depending upon the way that the product is taken by the patient;
  - For example, with oral products: *Candida* species, aflaxtoxin-producing *Aspergillus* species, *Bacillus cereus*, *Burkholderia cepacia*, Enterobacteriaceae (such as *Klebsiella* species), and other microorganisms were the population exceeds 100 CFU;
  - Some products carry a higher risk than others, for example, inhalation products and nasal sprays, optics, vaginal and rectal products, and oral solutions;
- The nature of the product. This involves considering whether the product supports growth, and if so whether it will support certain microorganisms more than others?

- The intended recipient: risk may differ for neonates, infants, the debilitated;
  - Limits for objectionable microorganisms in oral products intended for use by immunocompromised patient populations such as pediatric, human immunodeficiency virus (HIV), and cancer must be tighter than the limits for oral products intended for treating patients with diseases or conditions not affecting their immune systems because patients with deficient immune systems are more at risk of microbial infections [6];
- The presence of disease, wounds, organ damage;
- Where warranted, a risk-based assessment of the relevant factors is conducted by qualified personnel.

The list of objectionable microorganisms generated from a review should not remain static. The use of an experienced microbiologist to regularly screen and assess the microorganisms found in product and recovered from the environment needs to form part of an organization's continual risk assessment process.

To add to the complexity, a given microorganism may become "objectionable" under certain circumstances. The ways by which objectionable microorganisms trigger a risk to the product or have potential to cause patient harm include:

1. affecting product stability;
2. affecting the security of the container/closure system (is, for instance, the container adequately designed to retard access to the environment, and to prevent contamination from the environment?);
3. affecting the active ingredient;
4. producing off odors, flavors, or undesirable metabolites;
5. having the potential to grow and exceed the total aerobic count specification;
6. possessing high virulence and a low infective dose;
7. resistance to antimicrobial therapy.

When an objectionable microorganism is detected, this may or may not lead to rejection of the affected lot. To decide whether lot rejection must occur is dependent upon risk assessment. In risk assessing the impact of a microorganism on the product or process, a number of steps are required. These are [18]:

- Identity of the microorganism: find as much information as possible about the organism. This includes looking at recalls and other industries (e.g., medical and food) to determine if it could be a pathogen.
- Number of microorganisms present in the product: it is important to know the number of organisms present, especially when considering the infective dose. In addition, an assessment of the total number is important even if the microorganisms detected are not considered to be pathogenic. High numbers of nonpathogenic organisms may affect product efficacy and/or physical and chemical stability. It also stands that an unusually high number of microorganisms seen in the product may also indicate a problem during the manufacturing process;
- Microbial toxins. With this it is necessary to consider if the microorganism is likely to release a toxin (exotoxin, enterotoxin, or endotoxin) that could cause patient harm even if the microorganism is no longer viable.
- Consider the nature of the product: does it support growth? Does it have adequate preservation? Is it aqueous, cream, suspension, etc.?
- Assess the capability of the product to support growth or sustain the microorganisms. This requires knowledge of the inherent product characteristics, such as pH, water activity, and osmotic pressure. It should be noted that with the more resistant microorganisms, including

spore-forming bacteria, although they may not proliferate in a drug product with a low water activity, may persist within the product for long periods. In terms of some of the product characteristics:

- reduced water activity will greatly assist in the prevention of microbial proliferation in pharmaceutical products; and the formulation, manufacturing steps, and testing of nonsterile dosage forms should reflect this parameter [19];
- with pH, it should be assessed if the product pH in the same range as the ideal growth pH for the organism in question?
- with the product in general, it is important to assess of the product formulation contains ingredients that would be antimicrobial for the microorganism?
- With raw materials: consider processing to which the product is subjected, current testing technology and the availability of materials of desired quality.

In reviewing the outcome of the above evaluation: "is the microorganism objectionable?" and "when an objectionable microorganism is found, what is the risk?", a microorganism is likely to be classed both as objectionable and a high risk, if:

- the identification of the species has been confirmed;
- the patient population does not exclude those susceptible to the illness that this organism causes;
- the microorganism is known to cause illness;
- product route of administration is the same as the organism's route of infection (e.g., the bacterium causes illness via ingestion and the product is an oral product);
- the infective dose is low;
- it takes only a few cells to cause illness;
- it cannot be proven that the organism will not proliferate in the product.

With the above criteria, should they be met in whole or in part, then there would be little choice other than to reject the product.

8.4   Human microbiome project

With the establishment of the human microbiome project (HMP), knowledge of the diverse span of microbial species within and across the human body has been significantly enhanced, revealing valuable insight into community niche specialization, genetic diversity, and the prevalence of indigenous opportunistic pathogens. The HMP began in 2008 as a US National Institutes of Health initiative. The core objective of is to identify and characterize microorganisms associated with both healthy and diseased humans (the human microbiome) using a combination of culture techniques, metagenomics, and whole genome sequencing.

Arguably the outcomes from the analysis of the HMP have expanded the types of microorganisms that are considered to be objectionable. For example, upon exposure to pharmaceutical therapies containing antimicrobial preservatives, the diversity and composition of the human microbiome can be compromised, potentially resulting in physiological changes or the overgrowth of opportunistic pathogens. Conversely, the microbiome itself can also influence the human physiological response to pharmaceutical products, thus affecting the intended function of the product. A third consideration is that members of the human microbiome could be shed or deposited during the manufacturing process, thereby becoming an inadvertent source of contamination [20]. Thus, keeping abreast of the developments in this field is required in order for risk assessments in relation to objectionable microorganisms to be meaningful.

8.5   Conclusion

This chapter has presented a discussion about particular microorganisms and pharmaceutical products. This has centered on those microorganisms that are specified in the pharmacopeia as indicators of contamination and those that each facility must separately consider as "objectionable." Sometimes the self-assessed objectionables are the same as the compendial species and, at other times, they will be different.

The chapter has also presented approaches that can be taken for considering which organisms could be classed as objectionable and then, should such organisms be detected in a sample, how the impact of the detection can be risk assessed in terms of whether the material from which the sample was taken should be rejected.

This represents an important area since regulatory citations are relatively common in relation to objectionable microorganisms. Citations often center on the characterization of an objectionable microorganism in view of the product's intended use, the patient population (such as age and gender), patient health, dose, and application frequency of the medicine. Thus, the risk assessment process to enable assessments to be made needs to be scientifically sound and up-to-date.

References

[1]   USP. <62> Microbiological examination of nonsterile products: tests for specified microorganisms, USP 29, Supplement 2. Bethesda, MD: United States Pharmacopeia; 2006. p. 3765–9.

[2]   USP. <1111> Microbiological examination of nonsterile products: acceptance criteria for pharmaceutical preparations and substances for pharmaceutical use, USP 29, Supplement 2. Bethesda, MD: United States Pharmacopeia; 2006. p. 3801.

[3]   Jimenez L. Microbial diversity in pharmaceutical product recalls and environments. PDA J Pharm Sci Technol 2007;61(5):383–98.

[4]   Spooner DF. Hazards associated with the microbiological contamination of nonsterile pharmaceuticals, cosmetics and toiletries. In: Bloomfield SF, Baird R, Leak RE, Leech R, editors. Microbial quality assurance in pharmaceuticals, cosmetics and toiletries. Chichester, UK: Ellis Horwood; 1988. p. 15–34.

[5]   Charrock C. The microbial content of non-sterile pharmaceuticals distributed in Norway. J Hosp Infect 2004;57(3):233–40.

[6]   Manu-Tawiah W, Brescia BA, Montgomery ER. Setting threshold limits for the significance of objectionable microorganisms in oral pharmaceutical products. PDA J Pharm Sci Technol 2001;55(3):171–5.

[7]   Haas CN. Estimation of risk due to low doses of microorganisms: a comparison of alternative methodologies. Am J Epidemiol 1983;118(4):573–82.

 [8] Williams KP, Gillespie JJ, Sobral BWS, Nordberg EK, Snyder EE, Shallom JM, et al. Phylogeny of gammaproteobacteria. J Bacteriol 2010;192(9):2305–14.

 [9] Eckburg PB, Bik EM, Bernstein CN, Purdom E, Dethlefsen L, Sargent M, et al. Diversity of the human intestinal microbial flora. Science 2005;308(5728):1635–8.

[10] Fabrega A, Vila J. *Salmonella enterica* serovar typhimurium skills to succeed in the host: virulence and regulation. Clin Microbiol Rev 2013;26(2):308–41.

[11] Balcht A, Smith R. *Pseudomonas aeruginosa*: infections and treatment. New York: Informa Health Care; 1994, p. 83–4.

[12] Schneewind O, Fowler A, Faull KF. Structure of the cell wall anchor of surface proteins in *Staphylococcus aureus*. Science 1995;268(5207):103–6.

[13] Bartlett JG. *Clostridium difficile*: history of its role as an enteric pathogen and the current state of knowledge about the organism. Clin Infect Dis 1994;18:S265.

[14] Berman J, Sudbery PE. *Candida albicans*: a molecular revolution built on lessons from budding yeast. Nat Rev Genet 2002;3(12):918–30.

[15] De la Rosa M, Melum M, Vivar C. Microbiological quality of pharmaceutical raw materials. Pharm Acta Helv 1995;70(3):227–32.

[16] Sutton S. Microbial limits tests: the difference between "absence of objectionable microorganisms" and "absence of specified microorganisms". PMF Newsl 2006;12(6):3–9.

[17] Sutton SVW. What is an "objectionable organism"? Am Pharm Rev 2012;15:36–42.

[18] Wilder C, Sandle T, Sutton S. Implications of the human microbiome on pharmaceutical microbiology. Am Pharm Rev 2013;16(5):17–21.

[19] Werner D. Water activity: an underestimated parameter in pharmaceutical quality control. Pharmeuropa 2000;12(3):373–5.

[20] Turnbaugh PJ, Ley RE, Hamady M, Fraser-Liggett CM, Knight R, Gordon JI. The human microbiome project. Nature 2007;449(7164):804–10.

Microbial identification

9

9.1 Introduction

This chapter addresses some of the methods deployed for achieving species identification of an unknown microorganism (a bacterial or fungal species can be defined as a population of cells with similar characteristics). Hence, identification is the determination of whether a microorganism should be placed within a group of organisms known to fit within a classification scheme. Microbial identification can be defined as "microbial characterization by a limited spectrum of tests pre-chosen and appropriate to the problem being studied" [1].

Classically, microbial identification is undertaken using staining techniques and various agars and tests aimed at differentiating one probable species from another. This process was advanced during the 1970s through the advent of the API (analytical profile index) test strip, consisting of a series of militarized biochemical substrates contained within ampoules. Since the 1990s, a series of semiautomated phenotypic methods became available for the microbiology laboratory in a format that was relatively easy to use and relatively affordable [2]. The 2000s saw an equivalent range of more powerful genotypic methods introduced [3]. These different methods are discussed in this chapter.

9.2 Microbial taxonomy

The objective of microbial identification is to differentiate one microbial isolate from another and then to place that isolate into a family (genus) and a species (which is the best that can be achieved at the phenotypic level of identification) or even as a particular strain (through genotypic identification; a strain is a genetic variant or subtype of a microorganism). The differences between phenotypic and genotypic identification methods are outlined below.

In relation to taxonomy, this relates to the classification of an organism. The main taxonomic terms of importance to microbiology are:

- family: a group of related genera;
- genus: a group of related species;
- species: a group of related strains;
- type: sets of strain within a species (e.g., biotypes, serotypes);
- strain: one line or a single isolate of a particular species.

With an identification result, the most commonly used term is the species name (e.g., *Staphylococcus aureus*). There are always two parts to the species name, one defining the genus in this case "Staphylococcus" and the other the species (in this case "aureus"). Sometimes the species cannot be determined and the result,

Pharmaceutical Microbiology. http://dx.doi.org/10.1016/B978-0-08-100022-9.00009-8

drawing on the same example, would be *"Staphylococcus* species" (commonly the abbreviation "sp." is used in the singular or "spp." in the plural in place of the specific epithet. In this case, the microorganism is written in short-hand as *Staphylococcus* sp.).

9.3 Identification methods

Microbial identification is the determination of whether an organism should be placed within a group of organisms known to fit within some classification scheme. While it is possible for an experienced microbiologist to "identify" a microorganism by its visual appearance on a standard agar, such methods are generally unreliable and are no substitute for a standard identification method.

Identification methods can be divided into two groups: phenotypic and genotypic. The genotype–phenotype distinction is drawn in genetics. "Genotype" is an organism's full hereditary information, even if not expressed. "Phenotype" is an organism's actual observed properties, such as morphology, development or behavior [4]. The phenotype can alter, or at least appear different under varying environmental conditions. For example, a microbial colony may appear a different color on two different culture media.

Phenotypic methods are the most widespread due to their relatively lower costs for many laboratories. It should be recognized, however, that expressions of the microbial phenotype, that is, cell size and shape, sporulation, cellular composition; antigenicity, biochemical activity, and sensitivity to antimicrobial agents frequently depend on the media and growth conditions that have been used. In addition, phenotypic reactions typically incorporate reactions to different chemicals or different biochemical markers. These rely on the more subjective determinations. The reliance upon biochemical reactions and carbon utilization patterns introduces some disadvantages to the achievement of consistent (repeatable and reproducible) identification. To improve on the classical methods of biochemical identification, several developments have been made and refined in recent years. Collectively these methods are considered as modern biochemical identification techniques.

Genotypic methods are not reliant upon the isolation medium or growth characteristics of the microorganism. Genotypic methods have considerably enhanced databases of different types of microorganisms. Genotypic methods have opened up a whole new set of species and subspecies, as well as re-classifying species and related species (thus, taxa are often similarly grouped by phenotypic methods are actually polyphyletic groups, that is they contain organisms with different evolutionary histories which are homologously dissimilar organisms that have been grouped together) [5]. Genotypic methods utilize one of the two alternatives: hybridization or sequencing (most commonly of the gene coding for 16S rRNA (ribosomal ribonucleic acid)). With hybridization, DNA–DNA homology (or how well two strands of DNA from different bacteria bind (hybridize) together) is used to determine the relatedness of two microorganisms. With sequencing, the reason for methods examining the 16S rRNA region of the genome is:

- It is present in almost all bacteria, often existing as a multi-gene family, or operons;
- It is "highly conserved." The function of the *16S rRNA* gene over time has not changed, suggesting that random sequence changes are a more accurate measure of time (evolution);
- The *16S rRNA* gene is large enough for informatics purposes.

For filamentous fungal identification, this requires more expensive methods such as polymer chain reaction (PCR)-based internal transcribed spacer (ITS) regions sequencing by molecular methods. Advances have also been made with newer techniques, such as beta-D-glucandetection (using a (1–3)-β-D-glucanassay, based on the *Limulus* amebocyte lysate test) for the detection of fungal infections.

9.4 Phenotypic methods

Phenotypic methods allow the microbiologist to identify microbial species to the genus and sometimes to the species level based on a relatively small number of observations and tests. These are primarily growth-dependent methods, and identification must begin with a pure culture. The test comprises colony and cell morphology, Gram reaction and other staining characteristics, and metabolic and growth characteristics. The latter sets of tests are commercially available in test kits that are either read manually or through automated readers [6].

Before embarking on an identification test, the microbiologist needs to begin with a pure culture. Starting with a pure culture is the essence of good identification. This means that as a first step for identification is an aseptic sub-cultivation onto a suitable medium (certain media are required for specific microbial identifications systems), followed by incubation at a suitable temperature. Furthermore, with fungi, media will affect colony morphology and color, whether particular structures are formed or not, and may affect whether the fungus will even grow in culture. Therefore, the selection of media is as important as the subculture technique.

9.4.1 Colony and cell morphology

The first step of most identification schemes is to describe the colony and cellular morphology of the microorganism. Colony morphology is normally described by directly observing growth on agar, where the colony will appear as a particular shape (such as raised, crenated (having a scalloped edge), and spherical).

9.4.2 Staining techniques

9.4.2.1 Gram-stain

The primary staining technique used to differentiate bacteria is the Gram stain. The Gram stain is an important tool in the process of bacterial identification; this is through dividing bacteria into two groups (the so-called Gram-positives and Gram-negatives) and in allowing their morphological types (coccid or rod shaped) to be clearly seen by using a compound light microscope and oil immersion lenses (typically a 100× magnification is used).

The Gram stain method employed is a four-step technique: crystal violet, a tri-arylmethane dye (primary stain); iodine-potassium compound (mordant); alcohol or acetone (decolorizer); and safranin (counter stain). Carbol fuchsin is sometimes substituted for safranin since it more intensely stains anaerobic bacteria.

Iodine as the mordant means that it is a substance that increases the affinity of the cell for crystal violet so that crystal violet is more difficult to remove from the cell. With the test, Gram-positive organisms retain the crystal violet stain and appear blue; Gram-negative organisms lose the crystal violet stain and contain only the counter-stain safranin and thus appear red [7].

The chemical reaction at play is (Figures 9.1 and 9.2):

- *Step 1*: crystal violet (CV) dissociates in aqueous solutions to form CV+ and chloride (Cl⁻) ions. These ions penetrate through the cell wall and cell membrane of both Gram-positive

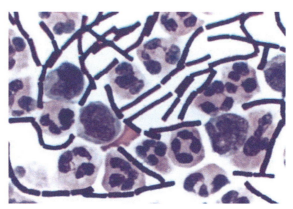

Figure 9.1 A Gram-positive stain, showing a *Bacillus* species.
Image: Creative Commons Library.

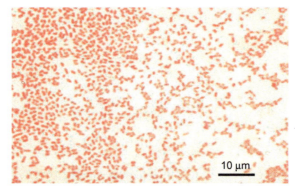

Figure 9.2 A Gram-negative stain, showing a *Pseudomonas* species.
Image: Creative Commons Library.

and Gram-negative bacteria. The CV+ ion interacts with negatively charged components of bacterial cells and stains the cells purple;

- *Step 2*: iodide (I⁻ or I⁻³) interacts with CV+ and forms large complexes of crystal violet and iodine (CV–I) within the inner and outer layers of the cell. Iodine acts as a trapping agent in that it prevents the removal of the CV–I complex;
- *Step 3*: when a decolorizer is added, it interacts with the lipids of the cell membrane. A Gram-negative cell loses its outer lipopolysaccharide membrane, and the inner peptidogly-can layer is left exposed. The CV–I complexes are washed from the Gram-negative cell along with the outer membrane.

In contrast, a Gram-positive cell becomes dehydrated from the decolorizer treatment. The large CV–I complexes become trapped within the Gram-positive cell due to the multilayered nature of the peptidoglycan;

- *Step 4*: after decolorization, the Gram-positive cell remains purple, and the Gram-negative cell loses its purple color. A counterstain (such as safranin) is applied to give decolorized Gram-negative bacteria a pink/red color.

Some bacteria, after staining with the Gram stain, yield a gram-variable pattern: a mix of pink and purple cells are seen. This can relate to the age of the culture (which is why cultures subcultured within 24h work best) or due to the nature of the bacterium (the genera *Actinomyces*, *Arthobacter*, *Corynebacterium*, *Mycobacterium*, and *Propionibacterium* have cell walls that are sensitive to breaking and, thus, some cells can appear "Gram-negative"; alternatively, with *Bacillus*, *Butyrivibrio*, and *Clostridium*, a decrease in peptidoglycan thickness during growth coincides with an increase in the number of cells that stain Gram-negative).

9.4.2.2 Bacterial spore stain

Physiological adaptation into the endospore production is a very important survival characteristic of some Gram-positive rods such as species of *Bacillus* and *Clostridium*. Endospore formation is usually triggered by a lack of nutrients; it is a stripped-down, dormant form to which the bacterium can reduce itself. The endospore consists of the bacterium's deoxyribonucleic acid (DNA), ribosomes, and large amounts of dipicolinic acid.

Spore staining using malachite green (a triarylmethane dye) and a safranin (an azonium compound) counterstain becomes a very useful tool in identifying the presence or absence of spores, and the location of spores such as terminal and subterminal, which may be used as a distinguishing feature in some spore formers. This is referred to as the Schaeffer–Fulton stain.

With this method, using an aseptic technique, bacteria are placed on a slide and heat fixed. The slide is then suspended over a water bath with porous paper over it, so that the slide is steamed. Malachite green is applied to the slide, which can penetrate the tough walls of the endospores, staining them green. After 5 min, the slide is removed from the steam, and the paper towel is removed. After cooling, the slide is rinsed with water for 30s. The slide is then stained with diluted safranin for 2 min, which stains most other microorganic bodies red or pink. The slide is rinsed again, and blotted dry and examined under a light microscope (Figure 9.3).

There are alternative staining methods, such as the Moeller stain, where carbol fuchsin (a mixture of phenol and basic fuchsin) is the primary stain used in this

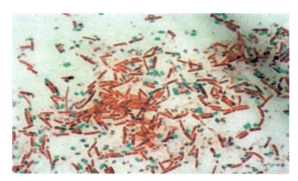

Figure 9.3 An endospore stain, where bacterial rods stain red and bacterial spores stain green. Image: Creative Commons Library.

method. Endospores are stained red, while the counterstain, methylene blue (a hetero-cyclic aromatic chemical compound) stains the vegetative bacteria blue.

9.4.2.3   Fungal staining

The identification of fungi using macroscopic and microscopic techniques is difficult and requires a trained eye. Lactophenol cotton blue stain is used in wet mounts for microscopically examining yeast and filamentous fungi. The stain serves as both a mounting fluid and stain. Staining the specimen light blue allows subtle features such as septa, special mycelia (hyphae weave together to form mycelium) and spore structures to be easily visualized by microscopy (Figure 9.4).

9.4.2.4   Ziehl–Neelsen stain

The Ziehl–Neelsen stain is a special bacteriological stain used to identify acid-fast organisms, mainly Mycobacteria. The reagents used are Ziehl–Neelsen carbol fuchsin, acid alcohol, and methylene blue. Acid-fast bacteria will be bright red after staining.

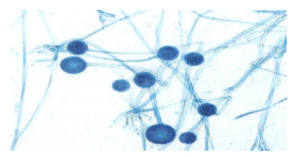

Figure 9.4 A fungal stain, using lactophenol cotton blue. Image: Creative Commons Library.

Given that, such bacteria are not readily detected within the pharmaceutical environment, and this stain is not discussed further.

9.4.3 Growth based and metabolic tests

Further identification examines the growth and metabolism of the bacterium. Differentiated culture media can be used here. This is media that selectively promotes the growth of certain bacteria. However, due to many variables of growth, such cultural techniques cannot always be assumed to be definitive. It is sounder practice to use a premade identification test kit.

The most common techniques used, based on their costs and long history, are biochemical tests. Biochemical test investigates the enzymatic activities of cells serve as powerful tests in the identification of bacteria. The basis of many biochemical tests is the fact that bacteria are capable of using different carbon sources to obtain the energy needed to sustain life. Which carbon sources react and which do not allows a probabilistic assessment to be made.

An example of biochemical profiling is the API identification system or the alternative BBL-crystal system (microtubes containing dehydrated substrates). The API was the first such identification test and was invented during the early 1970s by Pierre Janin of Analytab Products, Inc. (Figure 9.5).

Many laboratories now adopt semiautomated phenotypic identification systems, such as VITEK (a card preloaded with various biochemical broths) or OmniLog (a miniaturized system utilizing the microtiter plate format). Such phenotypic methods tend to work on the process of elimination. If test A is positive and B is not, then one group of possible microorganisms is included, and another is excluded. From this, tests C and D are performed, and so on. The test results are compared against databases that work on the basis of a dichotomous key [8]. An alternative is the analysis of cellular fatty acids by using gas chromatography (where patterns of fatty acid esters are determined by gas chromatography) [9]. Fatty acid methyl ester analysis by gas chromatography (GC-FAME) has been used for over many years to identify microbes in environmental and clinical settings; however, such systems are less common within the pharmaceutical microbiology laboratory.

More recently developed phenotypic methods include mass spectrometry and flow cytometry. Mass spectrometry can be orientated toward the identification and

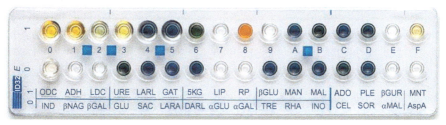

Figure 9.5 A 20E API identification test strip.
Image: Tim Sandle.

classification of microorganisms by using protein "fingerprints" (characteristic protein expression patterns that are stored and used as specific biomarker proteins for cross-matching). When identifying bacteria with a device like a Matrix-Assisted Laser Desorption Ionization Time-Of-Flight (MALDI-TOF) instrument, a single isolated colony or simple cell extract is spotted onto a stainless steel target plate and overlaid with an ultraviolet absorbing molecule. The target plate is inserted into the MALDI-TOF. Nitrogen-pulsed laser ionization is then applied to the sample, and the proteins are ionized. They are separated based on their mass/charge ratio. The resulting spectra, a protein fingerprint (which falls within the 2000–20,000 Da range), are compared with a database of known spectra.

Flow cytometry is a technique that employs serological methods to analyze cells suspended in a liquid medium by light, electrical conductivity or fluorescence as the cells individually pass through a small orifice. The use of fluorescent stains or fluorogenic substrates in combination with flow cytometry methods allows the detection and discrimination of viable culturable, viable nonculturable, and nonviable organisms [10].

9.5 Genotypic methods

Genotypic techniques study the microbial genome and, unlike phenotypic methods, they are not reliant upon the isolation medium or growth characteristics of the microorganism.

In bacteria, there are three genes that make up the rRNA functionality, these are: 5S, 16S, and 23S rRNA. Of these, the *16S rRNA* gene is most commonly used to identify the species. The 16S (small subunit) rRNA gene is selected for a number of reasons: (i) it is present in all organisms and performs the same function; (ii) its sequence is sufficiently conserved and contains regions of conserved, variable and hypervariable sequence; (iii) it is of sufficient size (around 1500 bases) to be relatively easily sequenced but large enough to contain sufficient information for identification and phylogenetic analysis (Figure 9.6).

An example of this technology is the RioPrinter (manufactured by Dupont Qualicon). This is an automated Southern blot device that uses a labeled ssDNA probe from the 16sRNA codon. The RiboPrinter uses a restriction enzyme, and strains can be identified and/or characterized by analyzing the ribosomal DNA banding pattern. Every time a sample is run, the RiboPrinter system produces an exact genetic snapshot of the microorganism that is linked to historical data. This genetic snapshot is akin to a "fingerprint." The DNA fingerprint is generated from regions of the rRNA genes (5S, 16S, 23S, and the spacer region including Glu-tRNA) that is unique to the microorganism at the strain level.

Another rapid method is a PCR system that uses a form of "bacterial barcodes" where the amplified genetic sequence is separated by gel electrophoresis and visualized to give a "barcode" specific to that strain. PCR is a technique which uses a DNA polymerase enzyme to make a huge number of copies of virtually any given piece of DNA or gene. It facilitates a short stretch of DNA (usually fewer than 3000 "base

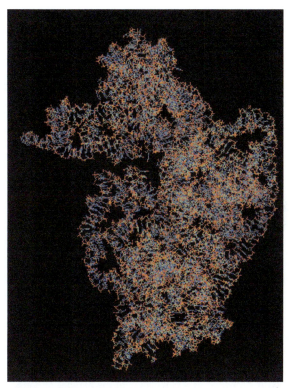

Figure 9.6 A representation of 16s rRNA.
Image: Creative Commons Library.

pairs") to be amplified by about one million-fold. With this comparative test, differences in the DNA base sequences between different organisms can be determined quantitatively, such that a phylogenetic tree can be constructed to illustrate probable evolutionary relatedness between the organisms [11]. An example of such a system is the MicroSeq manufactured by Applied Biosystems.

A final genotypic method is the Bacterial Barcodes system (DiversiLab). This system is also based on the PCR technology, using as a primer a sequence homologous to a repetitive sequence in the bacterial genome. The amplified sequence is then separated by gel electrophoresis and visualized to give the "barcode" specific to that strain.

9.6   Method validation

It is important that when microbiological identification methods are introduced into a laboratory that they are validated or verified. To begin with, if an automated instrument is purchased, this requires qualifying. Validation involves a series of steps, which can be summarized as:

1. Installation qualification: this is the documented evidence that the equipment and associated systems, such as software, hardware, and utilities, are properly installed, and relevant documentation is checked. Documentation may include manuals, certificates, procedures, and calibration records.
2. Operational qualification: this verifies that the system or subsystem performs as intended throughout all anticipated operating ranges and documents the information.
3. Performance qualification: this proves the system performs consistently as intended during normal operational use and remains in compliance with regulatory and user expectations or requirements. Performance of automated microbial identification system is very elaborate and time consuming due to multiple factors such as choice of isolates, operator variability, and the reproducibility of the system itself.

Following validation, or with nonautomated systems, verification of the test is required in order to show that it is suitable. Verification typically consists of [12]:

(a) parallel testing with approximately 50 microbial isolates using an existing system;
(b) the testing of 12–15 representative stock cultures of commonly isolates species (ensuring that these are of a broad enough range to cover the majority of the instruments test array). Here type strains should ideally be used;
(c) confirming that 20–50 microbial identifications, including 15–20 different species, agree with the results of a reference laboratory testing split sample.

The key criteria to be assessed are [13]:

(a) accuracy, which is expressed as a percentage of the number of correct results divided by the number of obtained results, multiplied by 100. To undertake this, type cultures from an approved culture collection should be used;
(b) reproducibility, which is similarly expressed as a percentage. Here the number of correct results in agreement is divided by the total number of results multiplied by 100;
(c) precision, which is achieved by testing multiple samples;
(d) ruggedness, which is undertaken by examining the same cultures with different reagents;
(e) robustness, which can be achieved through using cultures of different ages.

9.7 Conclusion

This chapter has outlined some of the microbial identification techniques undertaken. The techniques described have been divided between phenotypic and genotypic methods. It is important to note that groupings established by phenetic and phylogenetic systems do not always agree and within each grouping the methodological differences and varying contents of different databases will sometimes lead to conflicting analyses.

It is additionally important to understand that any systems used to identify bacteria, whether phenotypic or genotypic, will have limitations, because no single test methodology will provide results that are 100% accurate.

In terms of selecting between methods, this will depend on costs and resources, the time that the microbiologist is prepared to wait for and what level of identification is required. Some microbiologists are of the view that the only way to characterize a microorganism correctly is through a "polyphasic approach" that is a combination of phenotypic testing methods and genotypic testing methods. This is, however, far too time consuming and too prohibitively expensive for standard laboratories. Most routine testing laboratories select phenotypic test kits and use established contract test facilities where genotypic testing is required.

What is important, when making a selection, is to go back to basics and consider: what is the purpose of the identification? what does the microbiologist need to know? and what does the result tell the microbiologist? These questions can help with selecting and implementing the appropriate microbial identification test.

References

[1]   Stackebrandt E. Taxonomic considerations. In: Lewin RA, editor. Prochloron: a microbial enigma. USA: Chapman and Hall; 1989. p. 65–9.

[2]   O'Hara CM. Manual and automated instrumentation for identification of enterobacteriaceae and other aerobic Gram-negative bacilli. Clin Microbiol Rev 2005;18(1):147–62.

[3]   PDA. Evaluation, validation and implementation of new microbiological testing methods. Technical report no. 33. Parenteral Drug Association, 54(3), Supplement TR33; 2000.

[4]   Williams KP, Gillespie JJ, Sobral BWS, Nordberg EK, Snyder EE, Shallom JM, et al. Phylogeny of gammaproteobacteria. J Bacteriol 2010;192(9):2305–14.

[5]   Muller M, Davey H. Recent advances in the analysis of individual microbial cells. Cytometry A 2009;75(2):83–5.

[6]   Sutton SVW, Cundell AM. Microbial identification in the pharmaceutical industry. Pharmacopeial Forum 2004;35(5):1884–94.

[7]   Cowan ST, Steel KJ. Manual for the identification of medical bacteria. Cambridge, UK: Cambridge University Press; 1965.

[8]   Sandle T, Skinner K, Sandle J, Gebala B, Kothandaraman P. Evaluation of the GEN III OmniLog® ID system microbial identification system for the profiling of cleanroom bacteria. Eur J Parenter Pharm Sci 2013;18(2):44–50.

[9]   Fung DYC. Rapid methods and automation in microbiology. Compr Rev Food Sci Food Saf 2002;1:3–22.

[10] Attfield P, Gunasekera T, Boyd A, Deere D, Veal D. Application of flow cytometry to microbiology of food and beverage industries. Australas Biotechnol 1999;9:159–66.

[11] Kolbert CP, Persing DH. Ribosomal DNA sequencing as a tool for identification of bacterial pathogens. Curr Opin Microbiol 1999;2:299–305.

[12] Sandle T. Automated microbial identifications: a comparison of USP and EP approaches. Am Pharm Rev 2013;16(4):56–61.

[13] Green S, Randell C. Rapid microbiological methods explained. In: Halls N, editor. Microbiology and contamination control in pharmaceutical cleanrooms. Boca Raton, FL: CRC Press; 2004. p. 157–80.

Assessment of pharmaceutical water systems

10

10.1 Introduction

Water and the associated distribution systems can have a microbiological implications for both pharmaceuticals and healthcare [1]; this is especially so when water systems are poorly specified, improperly installed or not maintained to the appropriate standards. Microbiological risks are significant because water acts as a vector for microorganisms and it provides, with the addition of nutrients, the basis for microbial replication. Under most circumstances, the risks presented from water systems can be largely controlled by purification. This control is important for the use of water is, in pharmaceutical manufacturing, unavoidable.

Water is one of the most important materials within the pharmaceutical sector. Water is the basic ingredient of fermentation media, buffer manufacture, product extraction, purification, and as a solvent for dissolving products; furthermore, water is used for equipment cleaning, vial rinsing, diluting detergents, and so on [2]. While the quantities of water required will vary between facilities, one review estimated that 30,000 L of water is required to support the manufacture of 1 kg of a standard pharmaceutical [3]. Indeed the extensive use of water is one of the economic drivers for the adoption of single use, sterile disposable technologies. While disposable technologies can lead to economic savings through reducing the level of equipment washing and sterilization, the use of water as a pharmaceutical ingredient is inescapable.

Pharmaceutical facilities purify water through various treatments, culminating in either reverse osmosis or distillation. The objectives of water purification are threefold:

1. to reduce the levels of the chemical components in the water to prevent interactions with the drug substance, and to prevent toxicity;
2. to reduce the microbial bioburden to specified levels and to prevent further microbial proliferation;
3. to remove endotoxins and to prevent their future accumulation (this depends upon the grade of the water; not all grades of water are intended to the "endotoxin free").

The success of purification rests on the design and operation of the system. Due to the vagaries of these aspects, and based on the criticality of water in pharmaceutical production, water sampling and testing is subject to a high frequency of microbiological testing. This chapter examines the different types of water used in pharmaceutical facilities together with production methods. The chapter then proceeds to consider the microbiological risks and concludes with the main methods for microbiological monitoring.

Pharmaceutical Microbiology. http://dx.doi.org/10.1016/B978-0-08-100022-9.00010-4

10.2   Pharmaceutical facility water

The different types of water found in pharmaceutical production plants are potable water, purified water, highly purified water, and water for injection (WFI).

1. Potable water (sometimes called towns water or mains water).

This is water of drinking water standard, provided to the pharmaceutical company via the municipal water supply. Potable water is used for the routine cleaning of less critical areas, as with the preparation of detergents and disinfectants. It is also the source water for the purification step required to manufacture pharmaceutical grade water (purified water and water-for-injections) [4].

Private water companies or municipalities will supply potable water according to the local quality requirements. These requirements are designed to protect human health. Health protection is concerned with ensuring that levels of chemical pollutants remain within established safety criteria, and so that water-borne diseases will not be transmitted (such as the parasitic *Cryptosporidium* and bacterial pathogens). The types of microorganisms screened for include the indicator organisms: *Escherichia coli*, enterococci, and *Pseudomonas aeruginosa*.

The monitoring standard almost universally applied is a heterotrophic plate count, with a limit of 500 colony forming unit (CFU) per milliliter or less, and the absence of indicator microorganisms of fecal origin in samples of 100 mL. While testing is mandatorily conducted by the water providers, some pharmaceutical facilities elect to carry out parallel testing of the water provided to the site. Many pharmaceutical companies hold mains water in storage tanks, and it is prudent to sample this water to ensure that there is not an increase to the microbial levels.

2. Purified water

Purified water is used as a solvent in the manufacture of aqueous and oral products, such as cough mixture, and for the generation of fermentation products. It is also used in the preparation of detergents and disinfectants for the cleaning of certified cleanrooms of EU good manufacturing practice (GMP) Grade C/ISO 14644 class 8 (in operation) and those areas of a lower classification. Purified water also acts as the source of the steam supply to autoclaves. This grade of water is additionally used for the final rinsing of equipment and as the ingredient water for nonsterile products.

Purified water is typically produced by reverse osmosis. Reverse osmosis units use a semipermeable membrane and a substantial pressure differential to drive the water through the membrane in order to achieve chemical improvements, and microbial and endotoxin reductions.

Reverse osmosis systems exist in multiple design formats. In general terms, reverse osmosis functions as a size-excluding filter operating under a highly pressurized condition. An effective system will block 99.5% of endotoxin as well as ions and salts, while allowing water molecules through. In removing endotoxin, the system acts as a molecular sieve through which lipopolysaccharide cannot pass. The reader should be aware that there is some debate as to the relative effectiveness of reverse osmosis compared with distillation.

The microbial action limit is 100 CFU/mL (equivalent to 10,000 CFU/100 mL, given that the recommended test method is by membrane filtration).

3. Highly purified water

This is a type of water described in the European Pharmacopoeia (Ph. Eur.). The specification is very similar to the specifications for purified water, with the main difference being a specification for endotoxin. This grade of water is intended for use in the preparation of medicines where water of high biological quality is needed. It is used for sterile medicinal products that are not required to be apyrogenic such as, ophthalmic, nasal/ear, and cutaneous preparations. The water is prepared by reverse osmosis, and the microbial action limit is 10 CFU/100 mL. The endotoxin specification is set at the same level as per WFI (0.25 EU/mL).

4. Water-for-injection (WFI).

WFI is the "purest" form of pharmaceutical grade water. WFI is used for the generation of microbial fermentation media and the preparation of culture media use for cell lines. The water is also used as a raw material in the manufacture of pharmaceuticals intended to be sterile, and for the preparation of detergents and disinfectants used in higher grade cleanrooms, such as EU GMP Grade B/ISO 14644 class 7 (in operation) areas. Where water is required to reconstitute vials of lyophilized product, sterile WFI is provided (i.e., WFI that has been subject to a terminal sterilization process).

With WFI, the specifications of the Ph. Eur. and United States Pharmacopeia (USP) are very similar. However, there is a fundamental difference in opinion concerning its preparation. In the United States, WFI may be prepared either by reverse osmosis or by distillation, whereas the European authorities insist that only distillation be used for its production (there is, at the time of writing, some debate as to whether there will be harmonization with the US approach thereby permitting the use reverse osmosis. This does not have scientific consensus, and the main concern with reverse osmosis centers on the risk of endotoxin build-up).

Distillation functions by turning water from a liquid to a vapor and then from vapor back to liquid. Through this process, endotoxin is removed by the rapid boiling activity. This causes the water molecules to evaporate and the relatively larger lipopolysaccharide molecules to remain behind. Most models of distillation equipment are validated to achieve 2.5–3 log reductions in endotoxin concentration during distillation. This is based on lipopolysaccharide having a molecular weight of around 10^6 Da. Hence, endotoxin is heavy enough to be left behind when water is rapidly boiled off as in a still.

With the operation of distillation units, the principal concern is with the entrainment of contamination, particularly endotoxin. Low levels of Gram-negative microorganisms in the feed water will contribute endotoxin, which are concentrated by evaporation. In poorly designed or maintained systems, levels of endotoxin build-up can occur in the reservoir of the still.

To meet the requirements of the Ph. Eur., the microbial action limit is 10 CFU/100 mL, and the level of bacterial endotoxin must be <0.25 EU/mL. In order to show that the distillation unit is functioning as designed, it is good practice to monitor the endotoxin levels of the feed water to ensure that the challenge does not exceed

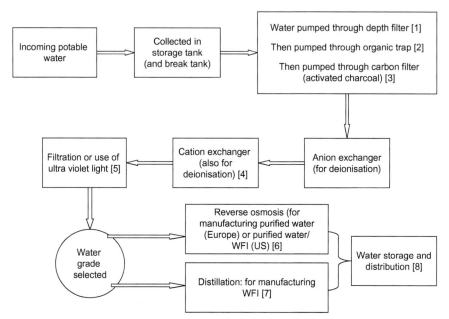

Figure 10.1 Diagrammatic representation of a pharmaceutical water system.

250 EU/mL (in order to generate WFI with an endotoxin level of less than 0.25 EU/mL, that is a three-log reduction has been achieved).

Figure 10.1 shows a typical schematic for the generation of pharmaceutical grade water.

In relation to Figure 10.1:

1. The depth filter is made from granular anthracite, washed sand, and gravel. The filter requires regular regeneration by backwashing;
2. The organic trap is resin, to which organic matter is removed from the water by binding;
3. The carbon filter absorbs residual organic materials, such as chlorine;
4. Anions, for example, Cl^- and SO_3^{2-}, are exchanged with hydroxyl (OH^-) counter ions from the anion exchanger;
5. Cations in water, such as Na^+, Mg^{2+}, and Ca^{2+}, are retained by displacing H^+ ions from the exchanger. Both the cation and anion exchangers are monitored by measuring restivity. As ions are exchanged and the process progresses, restivity should increase;
6. Filtration is performed using a 0.22 μm filter; ultraviolet light typically is of a wavelength of 254 nm;
7. Reverse osmosis uses a semipermeable membrane (made from polymers such as cellulose acetate or polyamides), which is permeable to water but not to microbial contaminants. Osmosis allows the movement of water across the membrane, moving from a solution of lower solute concentration to one of higher solute concentration, through the force of osmotic pressure. In some processes, double reverse osmosis is performed for additional assurance that microbial contaminants have been removed.
8. With distillation water is heated so that it condenses and is converted back into water; this process removes most impurities along with bacterial endotoxin.

9. The pharmaceutical grade water is held in a sealed storage vessel and fed around a network of pipes (distribution loop) to provide the manufacturing area. Unused water is returned to the storage vessel. The quality of the water is protected by a fast flow rate (typically greater than 9 ft/s; 2.7 m/s), which creates a constant, turbulent flow to minimize microbial attachment, and biofilm formation. Pipes are also designed to avoid dead-legs (to avoid stagnant water, as discussed below). To maintain the quality of the pipework, regular passivation and de-rouging are required. Furthermore, valves, especially those associated with user points, should be cleanable, free-draining, and not prone to leakage [5].

10.3   The microbial ecology of water

All water presents some form of microbial risk in that water provides a means for microorganisms to reproduce, and it is also an effective way of transferring microorganisms across distances. With the types of water found in the pharmaceutical facility, there are variations to the microbial ecology. With in-coming potable water, the microbial composition will vary depending on two factors. The first factor is the catchment area. Water used for the production of potable water is collected from vastly differing environments. These range from the nutrient-poor (oligotrophic) upland rivers where the microbial count, even using direct counting methods, will seldom exceed a few thousand CFU per milliliters, to the nutrient-rich (eutrophic) regions of lowland rivers, where counts can exceed 1 million per milliliters. The second factor is the season; the levels of microorganisms in natural waters follow a seasonal distribution curve controlled by the amount of available nutrients and temperature [6].

The composition of the microbial flora in the source water will be predominantly Gram-negative, containing prosthecate bacteria (bacteria that possess appendages), such as *Hyphomicrobium*, *Caulobacter*, *Gallionella*, and *Pseudomonas* species [7]. Bacteria are not the only microorganisms that inhabit source waters; the ecosystem will include fungi, protozoans, and algae. Contamination from land run-off and sewage contamination may add any number of potential pathogens [8].

With pharmaceutical water, through a process of increased purification, the complexity of the microbial ecosystem decreases as the diversity of the microorganisms prevalent within the water system correspondingly decreases. This is due to a reduction in niches within the ecosystem [9]. With purified water and WFI, the numbers of microorganisms should be very low. Generally, it is unlikely that the water generation will result in high levels of microorganisms—if the system is functioning well. Contamination tends to be introduced through poorly designed pipes or at user points.

10.4   Design and control of water systems

As indicated above, most microbial problems arise from the storage and distribution of the water rather than its generation. Primarily this arises through the development of biofilms, and, once established, biofilms can be extremely difficult to remove. In addition to distribution, improperly maintained generation components, such as carbon beds, softeners, reverse osmosis membranes (components illustrated in Figure 10.1), can also contribute to subsequent contamination downstream within the distribution system. Therefore, the design of water systems is of great importance.

An example of the importance of good design practices can be illustrated with WFI systems. WFI systems often stretch for hundreds of meters, sometimes across multiple floors; and the systems require high water flow rates [10]. Appropriate placement of outlets should be built into the design to assist operational logistics, to ensure representative sampling and the minimization of dead legs. As a part of biofilm control (as discussed below), stainless steel pipework or good quality plastic piping (such as polyvinylidene fluoride) will be used. In addition, sanitary design for valves is important. The design must minimize opportunities for water flow stagnation, for example, dead legs, or sites that allow for residue accumulation [11].

The requirements for well-designed water distribution pipes include:

(a) Smooth internal surfaces in tanks and in pipe-work. Microorganisms adhere less well to smooth surfaces than to rough surfaces. Pipe joints and welds can disrupt smoothness;
(b) Continuous movement of the water in tanks and rapid flow in pipework. Where shear forces are involved, microorganisms adhere poorly to surfaces. Where there is no movement of the water, there is no shear, shear increases with speed of flow;
(c) Avoidance of areas where water can remain stagnant.
   i. These include "dead legs"—water may stagnate in branch pipes branch from a circulating main if the length of the branch is too long to allow the turbulence of the flowing main to disturb the contents of the branch pipe. Here the principle is to always minimize the length of branch pipes;
   ii. Water can also remain stagnant in valves, particularly at user points and especially at user points which are not in frequent use. This problem can be counteracted by the use of so-called hygienic or "zero dead leg" valves. While these are significantly better than the alternatives (say ball valves), they should not lead to a false sense of security for they can harbor endotoxin-shedding biofilms;
   iii. Ring mains should be sloped (have "drop") from point of origin to the point of return to ensure that systems are completely drainable;
(d) Avoidance of leakage. Water leaks can cause bridging of water to the external environment through which bacteria may enter the system. Storage tanks should be equipped with filter on their air vents to prevent air-borne microbiological ingress. They may even be held under a "blanket" of an inert gas such as nitrogen;
(e) Controlled temperature storage and distribution is required for WFI systems. The risks of endotoxin-shedding biofilms despite the best attempts at control above are thought to be so consequential that the regulatory bodies require the temperature of storage and distribution to be maintained higher than 75 °C. It should, however, be considered that 75 °C is too high a temperature for most pharmaceutical formulation purposes. This means that user points are generally equipped with some form of cooling mechanism. It should be noted that heat exchangers used for this purpose may be a source of endotoxin and bacterial contamination and may thus cancel out many of the benefits of high temperature circulation.

In contrast to WFI, purified water systems can be maintained as a hot system or a cold system. Unlike WFI systems, purified water systems are typically cold systems and rely upon ultra-violet light and in-line filters to maintain microbial quality. Ultraviolet radiation (254 nm) is used for the disinfection of water of good optical clarity, and works particularly well in a re-circulating system where water flows over a multiple lamp system. If contamination occurs, both WFI and purified water systems should have the capability to be sanitized either by steam or by chemicals such as chlorine dioxide [12].

Even a good design can go wrong; therefore, operational control is also important. Operating procedures should require outlets to be flushed before usage to ensure use of the circulating water and to remove possible stagnant water or contamination from the surface of the outlet. Importantly, in the context of microbiological sampling, the flushing of outlets prior to sampling for monitoring purposes should be equivalent to that applied in operational use.

The use of hoses and temporary piping is a major source of contamination to product in during manufacturing, and therefore, their use should be minimized. Where used they should be subject to appropriate controls to minimize the risk of contamination from this source. For example, they should not be left on the outlets; they should be dried after use, hung vertically in appropriate locations to ensure free drainage, monitored, and, on a daily basis, cleaned, sanitized, and replaced.

With both the design and operation, the components of the storage and distribution system should enable validated cleaning and sanitization to be conducted.

10.5 Qualifying water systems

When a new water system is designed, it should be subject to formal qualification. Water systems require qualification based on bioburden, bacterial endotoxin, and organic and inorganic impurities [13]. Qualification steps include an operational qualification, where each outlet is monitored for a minimum of 2 weeks across each working day prior to the water being released for production; and a two-phase performance qualification. Phase I of the qualification will consist of 4 weeks of sampling as the water is being used by production, where samples should be taken at different times of production operations. Phase II is an assessment of the water over the course of 1 year. During the qualification, all out-of-limits results must be investigated and a root cause established.

10.6 Microbial contamination

Pharmaceutical water systems can become contaminated. Contamination can relate to special cause events, such as contamination at a user outlet (e.g., a hose left in a sink); and to common causes, which are systematic issues affecting the entire water system. A typical common cause can be caused by the formation of a biofilm.

10.6.1 Biofilms

Biofilms are made from a complex consortia of microorganisms organized within extensive exopolymer glycolices. Once formed, biofilms can be very difficult to remove, requiring a combination of heat and chemical treatments. The problems caused by biofilms include biofouling, biodeterioration, and physical blockage of industrial pipework and heat exchangers in water systems [14]. Microbial biofilms develop when microorganisms adhere to a surface by producing extracellular polymers that facilitate adhesion and provide a structural matrix. While the majority of bacteria are trapped within a biofilm, the biofilm will constantly generate bacteria that are released as free-floating individual cells and parts of the biofilm may slough off in clumps. As water is used and flows through the pipework or tap containing the biofilm, then the contamination risk arises at the point at which the water is used.

The steps involved in biofilm formation are:

1. individual cells populate the surface (initial attachment);
2. extrapolymeric substances (EPS) are produced, and attachment becomes irreversible;
3. biofilm architecture develops and matures;
4. single cells (or clumps of cells) are released from the biofilm.

The primary concern is that biofilms are highly recalcitrant and extremely difficult to remove once established.

Various design features may be utilized to prevent the development of biofilms and control contamination. The design of the system should include continual circulation with adequate flow rate to aid the prevention of biofilm formation (typically 1–3 m/s). Other common design features are the capacity to heat the water to elevated temperatures (as discussed above, elevating WFI to 75 °C or hotter) for sanitization purposes and the inclusion of high intensity ultraviolet (UV) light lamps. The inclusion of UV lamps downstream of potential microbial reservoirs, for example, carbon beds and softeners, has the added advantage of enabling ozone to be used for sanitization purposes. Where UV lamps are used, they should be regularly checked and maintained to ensure they are clean and provide the correct wavelength and energy output. The inclusion of filters within the distribution loop is difficult to justify and is not advisable.

Older systems, without such design features may require regular disinfection using an oxidizing agent to control biofilm (e.g., ozone, hydrogen peroxide, or hypochlorite). Each of these methods has disadvantages, including the need for additional design considerations (such as UV light to destroy ozone or extensive flushing is required to remove chemical residues if hypochlorite is used, which is expensive and disruptive [15]).

10.7   Microbiological sampling and testing

To assess the effectiveness of a water system, microbiological sampling is necessary. This includes taking samples of the incoming water at point of entry, the generation process used to produce water, the distribution tanks, and user outlets for both purified water and WFI. The frequency of sampling should be sufficiently high as to allow for meaningful trend analysis to be conducted. For this, at least some level of daily sampling is required (although each user point does not need to be sampled each day).

It is important that samples for microbiological analysis are taken appropriately. Good sampling practice includes taking samples through freshly autoclaved tubing, allowing outlets to be flushed for a standardized time or volume, and by taking the sample via good aseptic technique. Another aspect that is important is either testing samples within 2 h of the sample having been taken, or holding the samples at 2–8 °C prior to testing. The hold period, which would not ordinarily exceed 24 h, should be validated [16].

In terms of methodology, membrane filtration is the method of choice since a large volume of the sample is assessed. For potable water and water that is being processed through the generation plant (such as deionized water), some users elect to perform 1-mL plate counts. With microbiological agars, a standard plate count agar (PCA) is commonly used to assess potable water [17]. With purified water and WFI, the European Pharmacopoeia recommends Reasoner's 2A medium (R2A) agar. With the USP, no medium is recommended and the selection of the most appropriate culture medium rests with the site microbiologist (and ideally through a validation study). Nonetheless, the use of R2A is commonplace.

The widespread use of R2A is because it has long been recognized that total aerobic counts performed on water samples using low nutrient media (and preferably low 20–25 °C incubation temperatures) give much higher results in terms of microbial counts [18]. About 5–10-fold increases are often the norm when R2A is compared with general nutrient agar, such as tryptone soya agar [19]. The reason for this difference in magnitude is because bacteria undergo physical alterations and a metabolic downshift to survive in oligotrophic environments. Multiple genes are involved in this metabolic switching. This leads to bacteria in water being found in one of the two physiological conditions, and these conditions affect the ability of the bacteria to be recovered on different nutrient media.

Free-swimming bacteria (planktonic phase) are common to water systems, and it is these microorganisms that are collected and counted in water monitoring programs. The planktonic phase is energy expensive; it is a distribution strategy, but it is not a good survival strategy for starvation conditions. Survival mode is a switch from the planktonic phase to the benthic phase. Bacteria in the benthic phase lose motility, attach firmly to surfaces and start producing extracellular polymeric substances (EPS), which is the basis of a biofilm (as discussed earlier). EPS concentrate trace growth factors and afford protection from antagonistic agents such as biocides and heat treatments. Frequently reductions in cell size occur. Microorganisms in the benthic phase are often very difficult to culture on complex, rich media and they have also been described as viable but non-cultivable. Such bacteria can, however, be cultivated on low nutrient medium at lower incubation temperatures and extended (10–14 days) incubation times. Nonetheless, the value of a total aerobic count result obtained 14 days later is certainly questionable. Therefore, the Ph. Eur. "compromises" with the requirement for an incubation temperature of 30–35 °C and an incubation time of 5 days [20].

Even at 5 days, the value of an enumeration result is questionable, particularly so for water, which provides a dynamic environment. If the count exceeded an action level, it did so 5 days previously and possibly remained out of specification for the next 4 days. The net result is a difficult evaluation of the microbial quality of products manufactured during that period. This is why the emphasis should be on trend analysis rather than on individual results. It is also on this basis that there has been considerable investment and development in rapid and alternative microbiological methods (see below).

From this discussion about agars, it is important to appreciate that whatever cultural technique is used, it will only show a fraction of the microbial population in the sample. For that reason, the specifications for water counts are described as action limits; they are not considered to be pass/fail limits. If an action limit is exceeded, its impact on the product must be evaluated, but this does not often lead to batch rejection.

For WFI, water systems need to be assessed for bacterial endotoxins. This is undertaken through *Limulus* amebocyte lysate (LAL) methodology, with a limit of 0.25 EU/mL applied. LAL is discussed separately in this book. Although not directly related to microbiology, water systems are also examined for their chemical purity. Arguably, the most important examination is with total organic carbon (TOC) [21]. TOC is the amount of carbon bound in an organic compound and is often used as a nonspecific indicator of water quality or cleanliness of pharmaceutical manufacturing equipment. Although there is not a direct relationship with microorganisms, high levels of TOC may infer bacterial growth within the water system.

10.8 Action and alert limits

For the monitoring of water systems, appropriate alert and action levels should be set for both bioburden assessment and for levels of bacterial endotoxins. Action levels, where appropriate, are typically drawn from the pharmacopoeia or national water standards, whereas alert levels are assessed by pharmaceutical organizations, based on a review of historical data. The European and World Health Organization pharmacopoeial monographs for each type of water include statements on action limits; whereas the equivalent chapters within the USP recommend that "appropriate" monitoring limits be set. To set levels using historical data, ideally 1 year of data (or more) is analyzed in order to account for seasonal variations. There are different ways to calculate alert levels; one example is to take the 95th percentile.

Alert and action levels can be defined as follows:

- Alert limits are levels that when exceeded indicate that a process may have drifted from its normal operating condition. Alert limits constitute a warning and do not necessarily require a corrective action.
- Action limits are levels that when exceeded indicate that a process has drifted from its normal operating range. Exceeding an action limit indicates that corrective action should be taken to bring the process back into its normal operating range. Alert limits should be set below action limits.

If an action limit has been exceeded, the impact on the product(s) involved needs to be carefully investigated and evaluated. Furthermore, action and alert levels are useful markers for trend analysis. To assess the microbiology of water systems in a meaningful way, the collected data should be examined for trends; ideally on a monthly, quarterly, and annual basis (the latter allows an assessment of seasonality). Care must be taken when assessing microbial counts using traditional graphical tools, since such control charts are established on the basis that the data plotted are normally distributed. Microorganisms in water tend to follow Poisson distribution. Therefore, microbial counts require transforming prior to the results being plotted onto trend charts (such as by taking the square root or calculating the logarithm to base 10) [22].

When upward trends or action level excursions are recorded, investigations should be undertaken. The investigation should establish the cause of the excursion, and if possible, eliminate it. The evaluation should examine the impact on the product and its ability to withstand microbial challenge, as well as the patient group and their susceptibility to infection (this assessment requires identification of the contaminating microorganism). With action limit excursions, the investigation and evaluation should be carefully documented, and a justification for product release or rejection should be prepared. Some areas for investigation are displayed in Table 10.1.

Table 10.1   Table showing potential areas for the examination of water system problems in the event of microbial excursions

Main area	Areas for investigation
Sampling	Aseptic technique
	Adventitious contamination (type of microorganisms)
	Consumable/reagents/media—satisfactory?
	Were any pipes/valves leaking at time of sampling?
	Condition of sampling outlet
	Burst or leaking pipes
	Loss of pressure
	Identify who took the sample
	Staff training
	Integrity of container
	Interventions
	Transportation
	Storage
	Design of sample valve
	Check flushing
	Storage
Test method	Consumables—integrity/expiry date
	Reagents/media—storage/expiry date?
	Equipment—service/calibration
	Aseptic technique
	Test method—procedure followed?
	Incubation conditions
	Test controls
	Interpretation of result/calculations
	UDAF calibration/settle plates/airflow?
	Tubing—present/absent, does the tubing appear worn?

Continued

Table 10.1 **Continued**

Main area	Areas for investigation
Housekeeping/ specific outlet issues	Integrity of point: e.g., leaks, state of values, joints on tubing-outlet connections
	Past history of point reviewed
	Usage of point
	Temperature of outlet and/or sample
	Tubing—storage
	Steaming/sanitization satisfactory?
	Any issues for water supply?
Use	Establish what the water has been used for (e.g., direct product contact, such a dilution of product or buffer preparation?)/review usage of the point
	Any problems with plant samples?
Plant maintenance	Steaming/sanitization performed to set frequency?
	Review of plant history log
	Check flow rates
Design	Check loop temperature
	Check for dead legs
	Check valve design and maintenance
	Check task turnover rate
	Check task levels
	Check filters and change dates

10.9 Undesirable (objectionable) microorganisms

In addition to the examination of microbial counts, some facilities examine water systems of the presence of so-termed "objectionable microorganism." An "objectionable" microorganism is any microorganism that can cause infections when the drug product is used as directed or any microorganisms capable of growth in the drug product. In most situations, this can be translated to the absence of *P. aeruginosa* and *Burkholderia cepacia*, and the absence of any *Pseudomonas* spp. in nonsterile ophthalmic preparations. Occasionally, screening for *E. coli* is added as an indicator of fecal contamination (although it is unlikely that in-coming water is contaminated with such organisms). However, each pharmaceutical manufacturer must determine which microorganisms are classed as "objectionable" in relation to a specific process. The rationale of judgement will have to be based on product application and patient group vulnerability.

Examination for specified microorganisms requires the use of selective media and/ or enrichment steps; or, alternatively, specialized test kits. The recovery of suspect microorganisms from general test agar and identifying them is not acceptable. This is because the sensitivity of detection will be too low.

10.10   Rapid microbiological methods

Rapid (or alternative) microbiological methods have made some progress with water testing. The reason why such methods attract attention is not only a shorter time-to-result, but also because they detect a greater proportion of the microorganisms that are potentially present. If samples of water were tested by heterotrophic plate count and by direct counting methods (such as flow cytometry) the results in CFU/milliliter for the plate count would, in all probability, be in the range of 0.1–10% of the direct counts. This is because many of the microorganisms in water systems are unable to grow on plate count media; for some microorganisms, the media is too rich, and for other microorganisms, the cultural conditions are unsatisfactory. Improved recoveries are seen on R2A for prolonged periods at lower temperatures (as discussed above); however, the phenomenon of "viable but nonculturable microorganisms" means that many microorganisms found in water will not grow using cultural methods.

A range of rapid methods is available for the screening of water samples for indicators of contamination based on chromogenic, fluorogenic, or chemiluminogenic substrates. For example, with the examination of coliforms, the assays are based on the assumption that β-D-galactosidase and β-D-gluconidase are markers for coliforms and *E. coli*, respectively [23]. An alternative approach is with light scattering methods that can be used for the detection of water pathogens. With this method, as the slipstream passes through the flow cell, it also passes through a laser beam.

10.11   Microbiological assessment

The results from microbiological monitoring will typically be satisfactory over the course of 1 year. Sometimes over action results will be recorded from user outlets; more often these incidents are the result of poorly maintained sinks (where there is a risk of splashback) or through the local management of hoses. Such "special causes" events are rarely of a concern unless they occur in succession, as shown through a repeat sampling regime.

When systematic problems occur (including common causes such as biofilm formation), it is of great importance that microbiologists understand the basis of water system design. This is a key to root cause investigation. Concerns can also arise with the generation plant. With the production of pharmaceutical grade water, one of the weaknesses is that the resin beds can actually add microorganisms to the water if they are not properly maintained. A second concern is with the ion exchange process; here there is a risk if the ion exchange process does not remove microorganisms.

In order to gain sufficient oversight, the microbiologists should regularly review data from the water system and examine the data for trends. An example of a trend chart showing adverse drift is shown in Figure 10.2.

In Figure 10.2, the chart indicates the start of an adverse trend (emphasized by the addition of a linear plot trend line), thus an out-of-control situation, with a series of points rising above the upper control level. With the chart, the mean count has been

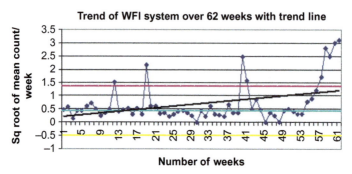

Figure 10.2 Plot of a microbial water system.

transformed through taking the square root of the mean count for each week. This is in order to approximate normal distribution.

In the example chart, the organization should have taken action concerning the water system and have closed the system down for investigation and formulation of appropriate preventative action. It is for this reason that adverse drift should always be investigated.

10.12 Summary

This chapter has examined pharmaceutical water systems. The chapter, in keeping with the theme of the book, has focused on the microbiological aspects of water systems. This concern with microbiological contamination has prevailed through considerations of system design, the risks presented from biofilms and with microbiological sampling. In examining these areas, it is clear that the site microbiologist should play an active role with the control and management of the water system. Much of this is based on a thorough examination of the data and in understanding how the production cycle impacts upon this, including high and low usage and seasonal variations. Water is of critical microbiological concern and it is important to be vigilant.

References

[1] Decker BK, Palmore TN. The role of water in healthcare-associated infections. Curr Opin Infect Dis 2013;26(4):345–51.

[2] Collentro WV. Pharmaceutical water: system design, operation and validation. 2nd ed. London: Informa Healthcare; 2011. p. 387–8.

[3] Walsh G. Biopharmaceuticals: biochemistry and biotechnology. 2nd ed. Chichester, UK: Wiley; 2003. p. 104–5.

[4] Sandle T, Saghee MR. Some considerations for the implementation of disposable technology and single-use systems in biopharmaceuticals. J Commer Biotechnol 2011;17(4):319–29.

[5] Collentro WV. Pharmaceutical water system design, operation, and validation. 2nd ed. London, UK: Informa Healthcare; 2011.

[6] Kawamua K. Qualification of water and air handling systems. In: Nash RA, Wachter AH, editors. Pharmaceutical process validation. 3rd ed. New York: Marcel Dekker; 2003. p. 401–42.

[7] Berry D, Xi C, Raskin L. Microbial ecology of drinking water distribution systems. Curr Opin Biotechnol 2006;17(3):297–302.

[8] Medema GJ, Payment P, Dufour A, Robertson W, Waite M, Hunter P, et al. Assessing microbial safety of drinking water improving approaches and method. In: Dufour A, Snozzi M, Koster W, Bartram J, Ronchi E, Fewtrell L, editors. WHO & OECD safe drinking water: an ongoing challenge. London, UK: IWA Publishing; 2003. p. 11–45.

[9] WHO (World Health Organization). Guidelines for drinking-water quality. 3rd ed. Recommendations, vol. 1. Geneva, Switzerland: WHO; 2008 [Incorporating 1st and 2nd Addenda].

[10] Cundell AM. Microbial monitoring of potable water and water for pharmaceutical purposes. In: Jimmenez L, editor. Microbiological contamination control in the pharmaceutical industry. New York: Marcel-Dekker; 2004. p. 45–75.

[11] Manfredi JJ. Water systems. In: Halls N, editor. Pharmaceutical contamination control. Bethesda, MD: PDA/DHI; 2007. p. 91–113.

[12] Sandle T. Avoiding contamination of water systems. Clin Serv J 2013;12(9):33–6.

[13] Haas CN, Karra SB. Kinetics of microbial inactivation by chlorine. Part II: kinetics in the process of chlorine demand. Water Res 1984;18:1451.

[14] Sandle T. Bacterial adhesion: an introduction. J Validation Technol 2013;19(2):1–10. On-line: http://www.ivtnetwork.com/article/bacterial-adhesion-introduction.

[15] Brown MRW, Gilbert P. Sensitivity of biofilm to antimicrobial agents. J Appl Bacteriol 1993;74:879–98 [Symposium Supplement].

[16] Sandle T. Water quality concerns: contamination control of hospital water systems. Eur Med Hyg 2013;(6):14–9.

[17] Sandle T, Roberts J, Skinner K. An examination of the sample hold times in the microbiological examination of water samples. Pharm Microbiol Forum Newsl 2009;15(2):2–7.

[18] Rosenfoldt EJ, Baeza C, Krappe DRU. Effect of free chlorine application on microbial quality of drinking water in chlorinated systems. J Am Water Works 2009;101(10):60–70.

[19] Reasoner D, Geldrich E. A new medium for the enumeration and subculture of bacteria from potable water. Appl Environ Microbiol 1985;49:1–7.

[20] Sandle T, Skinner K. Examination of the optimal cultural conditions for the microbiological analysis of a cold demineralized water system in a pharmaceutical manufacturing facility. Eur J Parenter Pharm Sci 2005;10(1):9–14.

[21] Hendricks DW. Water treatment unit processes: physical and chemical. Boca Raton, FL: CRC Press; 2007. p. 44–62.

[22] Sandle T. An approach for the reporting of microbiological results from water systems. PDA J Pharm Sci Technol 2004;58(4):231–7.

[23] Tryland I, Fiksdal L. Enzyme characteristics of beta-D-galactosidase- and beta-D-glucuronidase-positive bacteria and their interference in rapid methods for detection of waterborne coliforms and *Escherichia coli*. Appl Environ Microbiol 1998;64(3):1018–2013.

Endotoxin and pyrogen testing

11.1 Introduction

Bacterial endotoxin is the lipopolysaccharide (LPS) component of the cell wall of Gram-negative bacteria. It is pyrogenic, and it is a risk to patients who are administered intravenous and intramuscular preparations. This chapter outlines the pyrogenicity and the structure of endotoxin and moves onto examine *Limulus* amebocyte lysate (LAL) testing and other, alternative, methods of assessing pyrogens and endotoxin.

For these reasons, pharmaceutical products that are injected into the human body are tested for pyrogenic substances. The most common, and arguably most important pyrogen, is bacterial endotoxin. Bacteria endotoxin presents a significant risk to many pharmaceutical products, especially parenteral products. This is because endotoxin is [1]:

- ubiquitous in nature;
- has potent toxicity;
- is stable under extreme conditions;
- is likely to occur in the manufacturing process.

The pathological effects of endotoxin, when injected, are a rapid increase in core body temperature followed by extremely rapid and severe shock, often followed by death before the cause is even diagnosed. However, there needs to be large quantities of endotoxin within the human body the body for this to happen and it the endotoxin needs to be injected into the blood stream.

Of the available endotoxin tests, the LAL method is the most widely used [2]. The pharmacopoeial monographs for the LAL test (USP <85> and Ph. Eur. 2.6.14) are long established and relatively comprehensive and have been applied to the testing of parenteral products and water systems for bacterial endotoxin since the 1980s.

11.2 Pyrogenicity

Bacterial endotoxin can be classed, among other things, as a "pyrogen." Pyrogens are substances that, when injected into the mammalian body, will cause a variety of symptoms, the most recognizable of which is an increase in core body temperature. The association of fevers (pyrexia) has a long history. When injected into mammals (including humans), at a certain threshold, pyrogens will cause a number of adverse physiological responses. These responses include:

- increased body temperature;
- chilly sensation;
- cutaneous vasoconstriction;

Pharmaceutical Microbiology. http://dx.doi.org/10.1016/B978-0-08-100022-9.00011-6

- pupillary dilation;
- decrease in respiration;
- increase in arterial blood pressure;
- muscular pain;
- nausea and malaise.

Of these, a rise in body temperature represents the most common response, and the effect has been known since 1865, where it was reported that distilled water, later reasoned to be contaminated, triggered hyperthermia in dogs [3]. This physiological response is associated with the Greek word "pyrogen" (pyro, meaning "fire"; and gen signifying "beginning"). The term pyrogen was first used in 1876 [4].

In the early days of the pharmacopoeia, drug substances were classed as apyrogenic or pyrogenic, based, from 1942 and until the 1980s, on the "pyrogen test" (whereby a quantity of the drug was injected into three rabbits, and the temperature response of the rabbits was noted). The rabbit pyrogen test was first described by Florence Seibert in 1911, and it became a mainstay test for medicinal products from 1923 [5].

The rabbit test is no longer widely used, and it has largely been replaced, for the testing of parenteral drug products, by the LAL test. The reason for this is because the most common type of pyrogen found in the pharmaceutical industry is bacterial endotoxin, and for which, LAL (with some limitations explored below) is a specific test for. This risk from endotoxin is due, not least, to the large quantities of water used in the manufacture of pharmaceutical products as Chapter 10 describes. The predominance of the LAL assay is not to suggest that rabbit pyrogen testing has been completely eliminated, but that its use is in decline. Moreover, there are alternatives to the LAL test, such as enzyme-linked immunosorbent assay (ELISA) methods and the monocyte activation test (MAT). These are considered at the end of this chapter.

The LAL test is a method, of the bacterial endotoxin test (BET), for detecting the presence, and to go some way to determining the level, of Gram-negative bacterial endotoxins in a given sample or substance. Current editions of the pharmacopoeia carry statements to the effect that where the term apyrogenic or pyrogen-free is used it should be interpreted as meaning that samples of the product will comply with a limit for bacterial endotoxin.

However, it was not until the early twentieth century with the development of a rabbit pyrogen test that an understanding emerged in which bacteria could be classified into pyrogenic and nonpyrogenic types, correlatable to their Gram stain. Gram-negative bacteria were found to be pyrogenic, Gram-positive bacteria were generally not: and killed cultures of Gram-negative bacteria were comparable to live cultures in their ability to induce fevers [6]. Due to the association with living and dead bacteria, by the 1920s, it was apparent that sterility in parenteral pharmaceuticals could be no guarantee of nonpyrogenicity, and that if pyrogenicity was to be avoided it was imperative to avoid bacterial contamination at every stage of manufacture of parenteral pharmaceuticals.

In recognition that the causative agent of pyrogenicity was filterable and heat stable, efforts were applied to identifying its chemical composition. Trichloracetic acid and phenol–water extractions of bacteria were found to be effective in isolating the pyrogenic element from bacteria. These extracts were chemically identifiable as LPS (or what is commonly described as bacterial endotoxin). With pharmaceuticals, the greatest concern with pyrogens is with large volume parenterals. This is because these are products injected into the vein in relatively large volumes [7].

11.3   Bacterial endotoxin

The structural rigidity of the bacterial cell wall is conferred by a material called peptidoglycan (also known as murein). It is a polymer consisting of sugars and amino acids that forms a mesh-like layer outside the plasma membrane of bacteria forming the cell wall [8].

In Gram-positive bacteria, peptidoglycan is present as a thick layer that is outer-most in the cell wall. In Gram-negative bacteria, the peptidoglcan is only a thin layer, and it is not the outermost layer. Gram-negative bacteria instead have an outer membrane, and they are sometimes described as having a cell envelope rather than a cell wall. The outer membrane functions to maximize the ability of the bacterium to derive nutrients from the external environment. The outer layer also functions as a permeability barrier effective against diffusion of exo-enzymes into the external environment. This is an evolutionary feature that has arisen to allow Gram-negative bacteria (illustrated in Figure 11.1) to survive and increase in numbers in environments such as water in which there are only low concentrations of organic nutrients. Macromolecular organic nutrients are trapped in the cell envelope as the water flows by, and then within the cell envelope they are hydrolyzed to smaller molecules that can be taken into the cell.

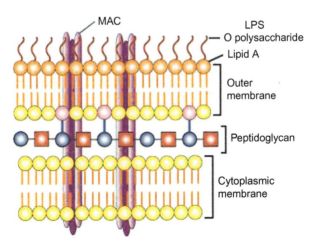

Figure 11.1 Gram's stain light microscope image, showing rod-shaped Gram-negative bacteria.
Image from Creative Commons Library.

The important evolutionary advantages conferred by LPS are:

- it contributes to adhesion of Gram-negative bacteria to surfaces allowing them to form as biofilms in aqueous environments;
- it "attracts" and "entraps" organic macromolecules from aqueous environments;
- it allows entrapped organic macromolecules to be "recognized" by the cell so that specific enzymes can be synthesized to break them down into smaller fragments capable of passing through the peptidoglycan layer into the cell;
- it increases the negative charge of the cell membrane and helps stabilize the overall membrane structure;
- it retains the enzymes synthesized by the cell, so that they are not lost into the external environment [9].

The outer membrane is composed of phospholipid and LPS protein. LPS is an amphiphile molecule—a hydrophilic polysaccharide attachment to a hydrophobic lipid moiety. LPS molecules vary in molecular weight from 1000 to 25,000 Da. LPS is rarely found as a unimolecular entity, it normally aggregates to form vesicles [10].

LPS is pyrogenic, and bacterial endotoxin is a synonym for LPS. LPS is the toxin which is synthesized endogenous to the bacteria cell structure. When Gram-negative bacteria are destroyed, endotoxin is released. In the human body, endotoxin triggers the activation of the body's defence system, which, in turn, elevates the body temperature and elicits the pyrogenic response [11]. LPS is located outside a thin structural layer of peptidoglycan, as shown in Figure 11.2.

Although intimately associated with the cell envelope of Gram-negative bacteria, LPS is constantly shed by the bacteria into the environment, much like the shedding of the outer layers of human skin. When Gram-negative bacteria die and lyze, all of their LPS is shed into the environment.

Furthermore, when bacterial cells are lysed by the immune system, fragments of membrane containing lipid A are released into the circulation, causing fever, diarrhoea,

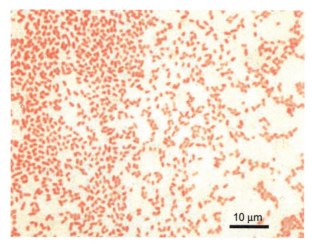

Figure 11.2 Diagram of the outer bacterial cell wall.
From Creative Commons Library.

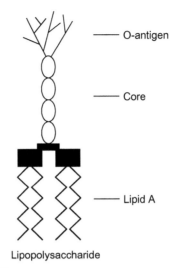

Figure 11.3 Diagram of LPS.
From Creative Commons Library.

and possible fatal endotoxic shock (also called septic shock) [12]. There are some other substances that are also pyrogenic. They are unusual and are extremely rarely found associated with pharmaceutical preparations.

LPS has three distinct chemical regions, as illustrated in Figure 11.3:

• an inner core called lipid A;
• an intermediate polysaccharide layer;
• an outer polysaccharide side chain.

Lipid A, embedded in the bacterial outer membrane, is responsible for pyrogenicity (endotoxic).

11.4 Quantifying endotoxin

An important assessment with any application of endotoxin testing is with endotoxin limits. Here, the universal unit of assessment is the "endotoxin unit" (EU). The approximate pyrogenic threshold dose for an adult human is 5 EU/kg/h, as set out in the pharmacopoeias. Any application of a product at a dose in excess of the 5 EU/kg/h will most likely result in a pyrogenic response. It has, therefore, been established that 5 EU/ kg/h is the maximum limit for most injectable products. It is also assumed that the mean adult human weight is 70 kg, and so the total amount of endotoxin that can be administered to a human per hour should not exceed 350 EU/h (70 kg × 5 EU = 360 EU/adult).

It is recognized that injectable products that are administered into the cerebrospinal fluid (intrathecal) require a much lower threshold limit of 0.2 EU/kg/h, or 14 EU/adult/h. It is important to consider the route of application when selecting the endotoxin limit. The term K is commonly used to represent the endotoxin limit. For a standard injectable product $K = 5\,EU/kg/h$, where intrathecally administered products K is $0.2\,EU/kg/h$.

The calculation of the endotoxin limit for a nonintrathecal, injectable drug product is illustrated in the following example:

Product A has a maximum human (whole body) dose of 1 g

$$\text{Dose per kg} = \frac{1\text{ g}}{70\text{ kg}} = 0.0143\,\text{g/kg} = 14.3\,\text{mg/kg}$$

$$\text{Endotoxin limit} = \frac{K}{M} = \frac{5\text{ EU/kg}}{14.3\text{ mg/kg}} = 0.35\text{ EU/mg}$$

The concentration of the product (potency in final product) is used to convert the endotoxin limit into EU/mL. If the concentration of product is $100\,\text{mg/mL}$, the following calculation will provide the EU/mL of the finished product. It is useful to use these units, as this is usually the units used within the LAL assay (as described below)

$$\text{Endotoxin limit} = 0.35\text{ EU/mg} \times 100\text{ mg/mL} = 35\text{ EU/mL}$$

It is important to note that this limit is based upon a specific product concentration. If the concentration of the product changes, then the endotoxin limit will change. This provides the maximum limit for this product at this concentration. It is common for manufacturers to select an endotoxin limit that is lower than this maximum limit.

The endotoxin limit of each sample requires determination prior to conducting endotoxin testing.

11.5 The *limulus* amebocyte lysate test

The most widespread endotoxin test is the LAL test. The principle of the LAL test is a reaction between LPS and a substance (clottable protein) contained within amoebocyte cells derived from the blood of the horseshoe crab, as illustrated in Figure 11.4 (of which *Limulus polyphemus* is the most commonly used species, although other species, such as *Carcinoscorpius* and *Tachypleus* demonstrate the same effect). The reaction of the horseshoe crab to endotoxin (the formation of a clot) has been known since the 1950s [13]. LAL is an aqueous extract obtained after lysis of blood cells (amoebocytes).

When endotoxin comes into contact with LAL, it initiates a series of enzymatic reactions that result in the activation of a pathway to the production of at least three serine protease zymogens (factor C, factor B, and pro-clotting enzyme). This pathway alters amoebocyte coagulogen (an invertebrate fibrinogen-like clottable protein) to form coagulin gel.

Serine proteases are enzymes that cleave peptide bonds in proteins, in which serine serves as the nucleophilic amino acid at the active site. They are found in humans as wells as in horseshoe crabs (and indeed in all mammals). In humans, they are responsible for co-ordinating various physiological functions, including digestion, immune

Figure 11.4 Image of the *Limulus* "horseshoe crab."
Image from Creative Commons Library.

response, blood coagulation, and reproduction. It is the blood coagulation reaction that is similar in both humans and horseshoe crabs [29]. It is on this basis that LAL has become the sensitive test reagent made from the soluble protein extract (lysate) of horseshoe crab blood cells (amoebocytes). All commercial lysates can detect picograms (10^{-12}g) of endotoxin.

The reference test in the pharmacopoeias is the gel clot (or gelation) and is conducted on the end-point principle. The description of the test and the necessary validation and accompanying controls is so detailed in both US Pharmacopeia (USP) and European Pharmacopoeia (Ph. Eur.) (harmonized since 1999).

The clotting mechanism of the blood of the crab is designed to prevent the spread of bacterial contamination throughout the horseshoe crab's biochemical system. When the endotoxin of Gram-negative bacteria contacts with the horseshoe crab's amoebocytes, a series of enzymatic reactions begin. The pathway alters amebocyte coagulogen into a fibrinogen-like clottable protein, which forms a coagulin gel. The defence mechanism is also effective against fungi, hence a similar reaction occurs in response to a fungal infection, which triggers the clotting cascade. In the reactions, β-glucans trigger the protease enzyme factor G, whereas endotoxin triggers the factor C enzyme, although the end result—coagulin—is the same [14].

A considerable amount more glucan (1000 times) is required to trigger the clotting cascade than the equivalent amount of endotoxin. The glucan required to trigger factor G can be of varying molecular weights (such weights range from 3 to 100 kDa) [15]. The clotting reaction is illustrated in Figure 11.5.

Glucan as an interfering substance is explored below.

The LAL reagent used for the gel-clot is supplied with an identified sensitivity or label claim (λ), for example, 0.03 EU/mL. This means that when mixed with an equal volume of the material under test, a gel or clot will form if the material contains 0.03 EU/mL or greater. For the kinetic methods, the lysate does not come with a label claim. The test sensitivity is determined by the lowest point of the standard curve used with each assay [16].

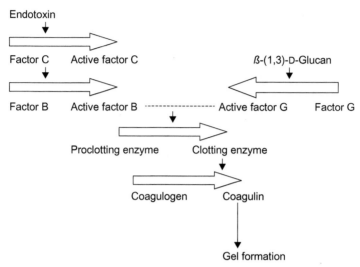

Figure 11.5 LAL-clotting cascade.
Adapted by the author.

When it is necessary to quantify endotoxin concentration in a material, it is usual to test a series of doubling dilutions against the reagent in temperature-controlled conditions. The greatest dilution that gives a positive result (formation of a gel or clot) is the end point, and the concentration of endotoxin in the original material can be calculated by multiplying the dilution factor at the end point by the sensitivity of the LAL reagent.

LAL tests require validation for each technician and each product before being applied routinely and, even then, valid assays require careful internal standardization. LAL tests require the use of standard endotoxin, termed a positive product control (PPC), which is a known amount of endotoxin mixed with a test material to confirm the absence of interference.

In the early days of the LAL test, endotoxin standards were variable. The potency varied with the method of purification, the bacteria of its origin and how it was formulated. Initially LAL test results were reported as units of weight. The problem with this was that results between tests and laboratories were not comparable. This was because different endotoxins, of the same weight and from different species of bacteria, can have different potencies. This led to the need for an endotoxin standard. This standard was based on *Escherichia coli* 0113: H10 K negative. This introduced the EU, which is a measure of the activity (potency) of endotoxin against the LAL test, irrespective of mass [17].

The units of measurement for the LAL test are EU. These are a measure of the activity of the endotoxin. Endotoxins differ in their biological activity or potency; the pyrogenicity or LAL reactivity of one endotoxin preparation may be very different from that of another of the same weight. Conversely, two endotoxin molecules may be different sizes and different weights but may have the same reactivity in an LAL test.

The potency of an endotoxin determined with one LAL reagent lot may differ from that determined with another lot. Expressing endotoxin concentrations in EUs avoids the issues of different potencies of different endotoxins and allows us to compare results of different LAL tests performed in different laboratories. Consequently, it is not the usual practice to convert results of an LAL test in EU/mL to units of weight of endotoxin per milliliter.

The origin of the test standard is from when the USP and US Food and Drug Administration (FDA) commissioned Rudbach to resolve the issue by preparing a reference endotoxin that was stable, could be chemically characterized, could be lyophilized without loss of activity and that was free from biologically active proteins [18]. This is termed reference standard endotoxin (RSE). Since RSE is expensive and potentially exhaustible, control standard endotoxin (CSE), also manufactured from *E. coli* 0113, is normally used for routine work but not necessarily for fundamental research. The potency of a given batch of CSE is determined relative to the current Ph. Eur., USP, or FDA lot of RSE and is specific to particular batches of LAL reagent. Thus, a purified form of LPS is used to manufacture a reference standard for the LAL assay. The endotoxin control is lyophilized with lactose and polyethylene glycol.

11.6   *Limulus* amebocyte lysate test methods

The primary LAL test methods are the gel-clot and two photometric methods: chromogenic and turbidimetric. The former is an end-point method while the latter two are kinetic methods. The kinetic methods are more sensitive than the gel-clot method (as low as 0.001 EU/mL). This is because changes in turbidity and color are discernible by light-scattering devices at lower concentrations of endotoxin than those at which gels form [19].

11.6.1   Gel-clot

Gel-clot is a method that utilizes the endotoxin-mediated clotting cascade, that naturally occurs within horseshoe crabs, to produce a gelatinous clot after incubation at $37 \pm 1\,°C$ with endotoxin. The gel-clot mimics the clotting of *Limulus* blood *in vivo*. Here a clotting protein is cleaved by an activated clotting enzyme, at which point the insoluble cleavage products coalesce and form a gel.

The gel-clot method is performed using depyrogenated glass tubes. The assay comprises of an equal volume of test solution and lysate (typically 0.1 mL of test solution and 0.1 mL of lysate) that are gently mixed together and, due to its sensitivity to vibration, incubated in an unstirred water bath or dry block heater at 37 °C normally for 1 h. After which, the end point is read by carefully inverting the tube through 180°. The tube is deemed positive for endotoxin if, after inversion through 180°, a solid clot (gelation) has formed and remains intact. The tube is deemed to be negative for endotoxin if no clot forms, or if the clot breaks through inversion of the tube.

11.6.2 Turbidimetric

The turbidimetric assay is, together with the chromogenic assay, a photometric method. With this test, during the process of clot formation, the lysate-sample reaction mixture becomes increasingly more turbid.

During the LAL reaction, the concentration of insoluble coagulin increases as its precursor, coagulogen, is cleaved. This process causes a corresponding increase in optical density (OD) of the reaction mixture. It is this increase in OD that is detected in a spectrophotometer (typically a microplate reader or a tube reader). The rate of increase of the OD is directly proportional to the endotoxin concentration present in the well. This is the basis for the turbidimetric LAL assay. The assay measures the increase in turbidity as a function of endotoxin concentration measured against a standard curve and, from this, estimates the endotoxin concentration in a sample.

11.6.3 Chromogenic

The chromogenic assay utilizes the initial portion of the endotoxin cascade. Here a synthetic chromogenic peptide is substituted for the clotting protein. The peptide generates a yellow color.

The chromogenic assay utilizes a synthetic chromogenic substrate that contains a specific sequence of amino acids that are designed to mimic the cleavage site in coagulogen. Activated clotting enzymes cleave this site and cause the liberation of the chromophore (para-nitro aniline, pNA), which has a yellow color. The liberated pNA absorbs light at 405 nm. In the chromogenic assay, the measurement of this absorption of light at this 405 nm that is measured. The degree and the rate of increased absorption is proportional to the endotoxin within the sample.

When using kinetic methods, the most important aspect is the standard curve for the endotoxin concentrations. This is because the standard curve selected, and how it performs, determines test sensitivity. Therefore, the high point and low point in a standard curve determines the lower and upper levels of endotoxin that can be detected.

Although the LAL test today is more robust, it remains open to a degree of variation [20].

11.7 *Limulus* amebocyte lysate test applications

As indicated above, the widest application of the LAL test method is with the testing of samples of water (primarily water-for-injection; WFI) and for assessing final products, especially those administered by injection. A related area is with depyrogenation studies.

WFI and sample testing are important because endotoxicity is not necessarily lost with the loss of viability of micro-organisms. LPS is not destroyed to any significant extent by sterilization treatments such as steam sterilization, gamma radiation, ethylene oxide, and hydrogen peroxide. LPS also passes through 0.22-µm bacteria-retentive filters. It is claimed that endotoxin may be removed from liquids by up to $4 \log_{10}$ reductions using 0.025-µm ultrafilters (which function as a molecular sieving process).

With depyrogenation, there are two commonly used ways of eliminating endotoxin from materials, removing them and inactivating them: by rinsing or by dry heat depyrogenation. The normal method of removal is by rinsing the material with WFI. This is normally applied to rubber stoppers for vials. It is also what is done to vessels and major pieces of equipment used for sterile parenteral manufacture. A question arises as to whether this can be validated and assured. In answer, the sampling statistics is likely to be poor, and the test method is inaccurate and probably there is not much there in the first place.

Inactivation is achieved by dry heat. If there are materials and glass vials that are required for sterile parenteral manufacture that can be depyrogenated by dry heat, then they should be depyrogenated by dry heat in an oven or a tunnel.

The regulatory standard for validation of an endotoxin inactivation (depyrogenation) process is that it should be capable of reducing an endotoxin challenge through $3\log_{10}$ reduction. To ensure that this limit works there is also a requirement to clean materials prior to dry heat depyrogenation with WFI—otherwise at least in theory, an item could be contaminated with 10,000 EU prior to entering a validated endotoxin inactivation process and still emerging with 10 EU intact and ready to contaminate the product.

Dry heat depyrogenation is a complex process that is still poorly understood with contradictory research data. The phenomenon that complicates the picture is that inactivation may approximate to the second-order chemical kinetics with a high initial rate of inactivation, then tail off to nothing. What this means in practice is that, at any particular depyrogenating temperature, it will be subjected to some degree of inactivation in some period of time or other but beyond that point no further inactivation will occur by holding the material at that temperature [21].

11.8   *Limulus* amebocyte lysate test interference

The LAL assay may be interfered with by the sample being tested; these are "LAL reactive materials (LRMs)" or "LAL activators." The phenomenon of substances reacting with LAL, such as thrombin and ribonuclease, has been known for many years (such as Elin and Wolff [28]). By "reacting," these LRMs can cause a positive LAL reaction for something other than endotoxin. This interference effect may be caused by a number of different factors such as pH, protein concentration, and presence of chemicals (such as NaOH from rinsing cycles). Interference may affect the lysate or the endotoxin. Inhibition or enhancement is normally detected through the use of spiked controls. Inhibition is arguably the greatest concern because it can result in a failure to detect the true level of endotoxin in a sample.

When conducting LAL tests, the dilution of samples is important. This is in order to minimize the effects of any component of the material being tested, which may be inhibitory to the LAL reaction. However, it is imperative that the material being tested is not diluted too much because this would ensure negative results, possibly false negatives. The way this is avoided is by having a maximum valid dilution (MVD), which is derived from the ratio of the endotoxin limit to the test sensitivity.

One potentially major source of interference is with β-glucans. These are polysaccharides of D-glucose monomers linked by β-glycosidic bonds. Glucans are important because they can react with certain lysates and cause interference with the LAL test [22]. Furthermore, glucans can, in certain circumstances, cause physiological effects in humans (β-glucans are known as "biological response modifiers" because of their ability to activate the immune system). According to Cooper et al., glucans are the most common LRMs likely to occur within pharmaceutical manufacturing [23]. Like endotoxin, glucans are large polysaccharides (homopolymers of glucose) of a high molecular weight. Therefore, the LAL test reacts with two types of polysaccharides: glucan and LPS (endotoxin). There are several sources of (1,3)-β-glucan. These include fermentation and cell culture media, fungal organisms, plant-derived materials, process unit operations' equipment, and packaging [24].

The extent to which glucans are a problem is a matter dependent upon the level of glucan and the patient population. Glucans are "biologically active" (immunostimulatory), although without the acutely toxic responses observed for similar levels of endotoxin. In some cases, glucans can cause a pro-inflammatory cytokine reaction. The main risk cited is for patients undergoing long-term haemodialysis. They cause a range of physiological reactions (such as the use of swabs in surgery) and are becoming increasingly regarded by regulators as a problem.

In response, most of the major LAL reagent manufacturers now produce lysate that is either glucan specific or endotoxin specific (in addition to the standard LAL reagent that could potentially react to either glucan or endotoxin). It should also be noted that the commercially available LAL reagents all differ in their reactivity to glucans (up to 200 fold in some studies, such as the experiments conducted by Roslansky) [25]. An alternative to modifications of the lysate is the use of a buffer, which can be added to the lysate to make it endotoxin specific. The buffer blocks factor G. Whether a glucan blocker should be used depends upon whether the testing laboratory wishes to know if glucan is present in the material under test or not.

LAL tests are only valid in situations where standard endotoxin can be shown to be detectable with the same efficiency in a test sample as in a control consisting of water (LAL reagent water, LRW) known to be endotoxin free (LRW).

11.9   Alternative test methods

There are alternatives to the LAL test. These are ELISA-based assays including the MAT.

MAT is an *in vitro* test used to detect or quantify substances that activate human monocytes or monocytic cells to release endogenous mediators. MAT works by predicting the human response to pyrogens on the basis of human fever, rather than animal models. MAT is based on human whole blood, and the test is theoretically capable of measuring all pyrogens relevant to the human patient (the MAT is based on the human fever reaction). In brief, the test uses heparinized human whole blood drawn from a healthy donor. The blood is diluted and introduced to the test sample. In response to any pyrogens present, monocytes contained within the blood sample will produce pro-inflammatory cytokines interleukin (IL)-$\beta1$ or IL-6, as measured by ELISA. These cytokines are pyrogens; in essence, all endogenous pyrogens are cytokines, molecules that are a part of the immune system. They are produced by activated immune cells and cause the increase in the thermoregulatory set point in the hypothalamus [26].

MAT is suitable for the testing of medical devices, blood products, toxic or immunomodulatory drugs, dialysis liquids, lipidic parenterals, and air quality [27]. In 2010, the MAT was accepted within Europe as an alternative endotoxin test method (Ph. Eur. 2.10).

An alternative to MAT is the endpoint fluorescent microplate assay called EndoLISA®. This revolutionary rapid method is based on an LPS-specific phage recombinant protein, which specifically binds to the entire substance group of LPS (endotoxin). The phage protein is precoated to the wells in the EndoLISA® microtitre plate, and as the sample is added to the well, the endotoxin (LPS) in the sample is bound to the phage protein. Any sample matrix with potentially interfering components is then completely removed by a washing step. Therefore, the subsequent detection by recombinant factor C (rFC) and a fluorescence substrate is left unaffected by inhibitors, facilitating a reliable quantification of endotoxin in the sample.

With the traditional LAL test, the industry appears to be moving towards the use of recombinant lysate, and the reliance upon the horseshoe crab may one day decrease. With recombinant, factor C in the clotting cascade has been replaced by a recombinant factor. Detection is by a fluorogenic substrate.

11.10   Conclusion

Pyrogens are a concern for pharmaceutical drug products and for many of the ingredients used to formulate them. This is especially so for products that come into contact with human blood. Here, by far the most concerning pyrogen is bacterial endotoxin. In relation to this, the chapter has considered the risks of endotoxin to pharmaceutical processing and some of the control measures in place to reduce the risk of endotoxin contamination.

Furthermore, the chapter has provided an introduction to endotoxin as well as to the primary method for detecting endotoxin: the LAL test. Here the chapter has examined the three main types of LAL test: gel-clot, chromogenic, and turbidimetric, as well as considering alternative test methods. Such tests are an essential feature of most pharmaceutical microbiology laboratories.

References

[1] Williams KL. Endotoxin: relevance and control in parenteral manufacturing. In: Jimmenez L, editor. Microbiological contamination control in the pharmaceutical industry. New York: Marcel-Dekker; 2004. p. 183–249.

[2] Guy D. Endotoxins and depyrogenation. In: Hanlon G, Hodges N, editors. Industrial pharmaceutical microbiology: standards and controls. Passfield, UK: Euromed; 2003. p. 12.1–12.15.

[3] Billroth T. Arch Klin Chris 1865;6:372.

[4] Sanderson JB. On the process of fever. Practitioner 1876;16:257.

[5] Seibert FB. The cause of many febrile reactions following intravenous injections. Am J Physiol 1923;71:621.

[6] CoTui HD, Schift MH. Production of pyrogen by some bacteria. J Lab Clin Med 1942;27:569.

[7] Akers MJ, Larrimore DS, Guazzo DM. Parenteral quality control: sterility, pyrogens, particulate, and package integrity testing. 3rd ed. New York: Marcel Dekker; 2003. p. 119–20.

[8] Madigan MT, Martinko JM, Dunlap PV, Clark DP. Brock biology of microorganisms. 12th ed. San Francisco, CA: Pearson/Benjamin Cummings; 2009.

[9] Raetz CRH, Whitfield C. Lipopolysaccharide endotoxins. Annu Rev Biochem 2002;71:635–700.

[10] Gould MJ. *Limulus* amebocyte lysate assays and filter applications. In: Jornitz MW, Meltzer TH, editors. Filtration and purification in the biopharmaceutical industry. New York: Informa Healthcare; 2008. p. 425–38.

[11] Gould MJ. Evaluation of microbial/endotoxin contamination using the LAL test. Ultrapure Water 1993;10(6):43–7.

[12] Brandberg K, Seydel U, Schromm AB, Loppnow H, Kochm MHJ, Rietschel ET. Conformation of lipid A, the endotoxin center of bacterial lipopolysaccharide. J Endotoxin Res 1996;3(3):173–8.

[13] Bang FB. A bacterial disease of *Limulus polyphemus*. Bull Johns Hopkins Hosp 1956;98:325.

[14] Moser K. Playing hide and seek with endotoxin. LAL User Group Newsl 2009;2(3):1–5.

[15] Tanaka S, Aketagawa J, Takahashi S, Shibata Y, Tsumuraya Y, Hashimoto Y. Inhibition of high-molecular-weight-(1β3)-β-D-glucan-dependent activation of limulus coagulation factor G by laminaran oligosaccharides and curdlan degradation products. Carbohydr Res 1993;244:115–27.

[16] Upton A, Sandle T. Best practices for the bacterial endotoxin test: a guide to the LAL assay. Stanstead Abbotts: Pharmaceutical Microbiology Interest Group; 2012.

[17] Dawson ME. Endotoxin testing. In: Halls N, editor. Pharmaceutical contamination control. Bethesda, MD: PDA/DHI; 2007. p. 146–86.

[18] Rudbach JA, Akiya FI, Elin RJ, Hochstein HD, Luoma MK, Milner CB, et al. Preparation and properties of a national reference endotoxin. J Clin Microbiol 1979;3(10):21–5.

[19] Berzofsky R. Endotoxin, limulus amebocyte lysate and filter applications. In: Jornitz MW, Meltzer TH, editors. Filtration and purification in the biopharmaceutical industry. New York: Informa Healthcare; 2008. p. 413–23.

[20] McCullogh KC, Weider-Loeven C. Variability in the LAL test: comparison of three kinetic methods for the testing of pharmaceutical products. J Parenter Sci Technol 1992;46(3):69–72.

[21] Sandle T. Sterility, sterilisation and sterility assurance for pharmaceuticals: technology, validation and current regulations. Cambridge: Woodhead Publishing; 2013. p. 171–188.

[22] Sandle T. Pharmaceutical product impurities: considering beta glucans. Am Pharm Rev 2013;16(5):16–9 [Special Edition Supplement 'Furthering Pharmaceutical Microbiology'].

[23] Cooper JF, Weary ME, Jordan FT. The impact of non-endotoxin LAL-reactive materials on limulus amebocyte lysate analyses. PDA J Pharm Sci Technol 1997;51(1):2–6.

[24] Pearson FC, Bohon J, Lee W, Bruszer G, Sagona M, Jakubowski G, et al. Characterization of limulus amoebocyte lysate-reactive material from hollow-fiber dialyzers. Appl Environ Microbiol 1984;48:1189–96.

[25] Roslansky PF, Novitsky TJ. Sensitivity of *Limulus* amebocyte lysate (LAL) to LAL-reactive glucans. J Clin Microbiol 1991;29(11):2477–83.

[26] Dinarello CA. Proinflammatory cytokines. Chest 2000;118(2):503–8.

[27] Schindler S, von Aulock S, Daneshian M, Hartung T. Development, validation and applications of the monocyte activation test for pyrogens based on human whole blood. ALTEX 2009;26(4):265–77.

[28] Ellin RJ, Wolff SM. The effect of pH and iron concentration on the growth of *Candida albicans* in human serum. J Infect Dis 1973;127:705–08.

[29] Hellum M, Øvstebø R, Brusletto BS, Berg JP, Brandtzaeg P, Henriksson CE. Microparticle-associated tissue factor activity correlates with plasma levels of bacterial lipopolysaccharides in meningococcal septic shock. Thromb Res 2014;133(3):507–14.

Sterilization and sterility assurance

12

12.1 Introduction

Many types of pharmaceutical products are required to be sterile. This includes injections, infusions, and pharmaceutical forms for application on eyes and on mucous membranes [1]. These methods of administration are often due to the formulation of the products (e.g., the active ingredient might be inactivated if they were to be ingested). Because of the route of administration, such medicines are required to be sterile. Importantly, if such medicines are not sterile, then this could lead to patient harm or even death.

Manufacturing sterile products not only requires the product solution to be sterile. Sterility must encompass the various components required for the production and development of sterile products. Safeguarding the patient not only extends to manufacturing a sterile product, it needs to include the use of sterile items to administer the drug (such as a sterile syringe and needle) and administering the drug under aseptic conditions, using trained medical or nursing practitioners.

Thus, sterile manufacturing itself is a continuum that stretches from development to manufacturing, to the finished product, to marketing and distribution, and to utilization of drugs and biologicals in hospitals, as well as in patients' homes. There is no generic approach to the manufacturing of sterile products. Each plant or process will differ in relation to the technologies, products, and process steps. The common point is that a product is produced that is sterile and where there is no risk of contamination until the contents of the outer packaging are breached (such as through the injection of a needle through a bung on a product vial).

12.2 Sterility

Sterility can be defined as "the absence of all viable microorganisms." Therefore, something would be deemed sterile only when there is complete the absence of viable microorganisms from it. Sterility is an absolute term. Either something is sterile or it is not. There is no such thing as "slightly sterile" or "almost sterile."

Following on from this, sterilization can be taken to mean the use of a physical or chemical procedure to destroy all microbial life, including highly resistant bacterial endospores. This destruction of bacterial spores means that sterilization is a complete process for the destruction of life, unlike disinfection, which refers to the reduction of a microbial population by destruction or inactivation.

Importantly, this simple definition refers to microorganisms that are "viable" (that is bacteria, fungi, and viruses that are capable of reproducing under the correct conditions). It does not, however, refer to the absence of microbial by-products. By-products

Pharmaceutical Microbiology. http://dx.doi.org/10.1016/B978-0-08-100022-9.00012-8

include toxins that may cause harm, such as endotoxins, exotoxins, or enterotoxins. These can be released by microorganisms as the function or when they die and several toxins are resistant to many types of sterilization (e.g., for endotoxins, a depyrogenation process is required).

Furthermore, the term "sterile" does not extend to other aspects of the formulation, which might cause patient harm, such as the presence of particulates or chemical impurities. Moreover, something that is sterile, such as a liquid in a bottle, is only sterile at a point in time; something that has been rendered sterile can become nonsterile if there is ingress of microorganisms (such as a crack in the bottle leading to microbial ingress). Thus, something that has been rendered sterile is subject to the possibility of becoming nonsterile under certain conditions [2].

Although the definition of sterility—"the absence of all viable microorganisms"— is straightforward, the evidence that something is sterile can only be considered in terms of probability. This is because absolute sterility can only be proved by testing every single item produced (and with technology that will give an undisputable result). However, the act of testing destroys the very item that is required for administration to the patient, so sterility cannot be proven empirically.

Therefore, the concept of what constitutes "sterile" is measured as a probability of sterility for each item to be sterilized [3]. Probability can be considered in relation to components that are sterilized and to products that can be terminally sterilized in relation to a concept called the sterility assurance level (SAL). Importantly, the SAL concept cannot be applied to aseptically filled products. With aseptically filled products, the probability of sterility or nonsterility is the product of environmental controls (from clean air devices and cleanrooms), product filtration, the use of sterilized components, personnel behaviors, and gowning.

12.3   Sterility assurance and the sterility assurance level

The manufacture of sterile products involves the philosophy and application of sterility assurance. Sterility assurance, as a broad term, refers to the philosophy of protecting a sterile product throughout its manufacturing life in relation to controls and practices. It is not synonymous with the SAL, although the reduction of the two concepts is, unfortunately, too common. The term "sterility assurance" is a combination of two words with the following definitions:

- sterility—state of being free from viable microorganisms;
- assurance—a positive declaration intended to give confidence.

Sterility assurance concerns the wider embracement of the aspects of good manufacturing practice (GMP) that are designed to protect the product from contamination at all stages of manufacturing (from in-coming raw materials through to finished products) and, thus, it forms an integral part of the quality assurance system.

A quantitative assessment of the sterility assurance can be provided through the SAL, a term used to describe the probability of a single unit being nonsterile after a batch has been subjected to the sterilization process (or the probability of a single viable microorganism surviving on or in an item after sterilization) [4]. Importantly, the SAL concept was developed for sterilization processes, and it should be limited to terminal sterilization; thus, it cannot, as a probabilistic concept, be applied to aseptic manufacture (while certain literature attempts to do so, such attempts should be avoided for they are scientifically inaccurate).

A second important point is that the SAL is not exactly a definition of the assurance of "sterility"; rather it is the probability of "nonsterility" [5]. SALs are used to describe the probability that a given sterilization process has *failed* to destroy all of the microorganisms. This is why the term is defined as the probability of a treated item remaining contaminated by one or more viable microorganisms and not, as sometimes misreported, the probability of successful sterilization.

The reason that sterilization is discussed in terms of probability is because it is impossible to prove that all microorganisms have been destroyed. This is because:

(1) microorganisms could be present but undetectable simply because they are not being incubated in their preferred environment;
(2) microorganisms could be present but undetectable because their existence has never been discovered.

SALs can be used to estimate the microbial population that was destroyed by the sterilization process. Each log reduction (10^{-1}) represents a 90% reduction in microbial population. So a process shown to achieve a "6-log reduction" (10^{-6}) will reduce a population from a million microorganisms (10^6) to very close to zero (theoretically). The same logic can apply to containers as to microorganisms. For example, a SAL of 10^{-6} expresses probability of survival, that is, there is one chance in 10^6 that any particular container out of 10^6 containers would theoretically not be sterilized by the process [6].

SAL is demonstrated through validation using innocuous bacterial endospores (biological indicators). The assumption is that the inactivation of such highly resistant microorganisms encompasses all less-resistant organisms, including most pathogens [7].

The concept is also linked to the predictability of microbial death through the use of a defined sterilization process. Microorganisms die when exposed to a sterilization treatment according to a logarithmic relationship, which is based on the proportion of viable microbial cells and the time of exposure to the sterilization process (such as heat) or to the dose (such as radiation).

In assigning a quantitative value a SAL of 10^{-6} takes a lower value but provides a greater assurance of sterility than a SAL of 10^{-3} [8]. Furthermore, the SAL is normally expressed as 10^{-n}. For example, if the probability of a spore surviving was one in one million, the SAL would be 10^{-6}. The reader will note that the SAL is a fraction of one and, therefore, carries a negative exponent (so the six-log reduction is written as 10^{-6} rather than 10^6). However, the reader should be aware that SAL refers to individual items of product and not to a batch of product (Figure 12.1).

This theoretical reduction in microbial population also assumes [9]:

· a single species of microorganism present on or in each product;
· there is a homogenous microbial population;

Figure 12.1 Biological indicators used to determine the sterility assurance level of an autoclave. Photograph: Tim Sandle.

- the population has a mono-disperse distribution on the surfaces to be sterilized, that is, there is no clumping;
- the exclusion of multinucleate spores (e.g., ascospores) or microorganisms.

For many years, a SAL of 10^{-6} has represented the sterilization standard for invasive and implantable devices and medicinal products administered by injection. In practice, many processes use "overkill cycles," which assure an even lower probability that a device will be nonsterile [10].

12.4 Sterility testing

A common means to assess the effectiveness of sterility for medicinal products is, the sterility test. Sterility testing is described in detail in Chapter 6.

Sterility testing is less common for the sterilization of consumables where a terminal sterilization process is used (here the product bioburden is assessed prior to sterilization and compared to validation cycles for the microorganisms found in relation to the population and resistance of the microorganisms to the sterilization process). For medicinal products that can be terminally sterilized, parametric release is often used in lieu of the sterility test (see below). For aseptically filled products, and some terminally sterilized products, the sterility test is a regulatory requirement (as per the US Food and Drug Administration (FDA) and the European Medicines Agency).

Despite the requirement to conduct the sterility test on a representative batch size, it remains that the sterility test is a flawed test on a number of levels. The first relates to the very small sample size tested. Testing any pharmaceuticals and medical devices to a level of statistical significance would require a sample size that would be practically and economically unsustainable. Second, the microbial challenge to the manufacture of pharmaceuticals and medical devices includes microorganisms that, by virtue of their fastidious nature, or physiological prerogative, will not grow on growth medium.

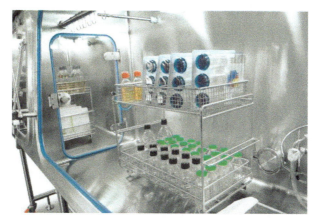

Figure 12.2 Isolator prepared for sterility testing.
Photograph courtesy of Pharmig.

There is a evidence that microorganisms in forms that will not necessarily replicate still retain their ability to cause disease [11].

Validation of the sterility test involves challenging the test with a small number of known microorganisms. Validation of the sterility test is important in order to show that the culture media and the conditions used during the test neutralize any antimicrobial activity that the product possesses, and that microorganisms can be recovered (Figure 12.2).

12.5 Parametric release

Products that can be terminally sterilized can be subject to parametric release without undertaking finished product testing. The European Organization for Quality defines parametric release as: "A system of release that gives the assurance that the product is of the intended quality based on information collected during the manufacturing process and on the compliance with specific GMP requirements related to Parametric Release." Importantly, the organization must demonstrate the capability of the sterilization agent to penetrate to all relevant parts of the product [12].

Parametric release assumes that a robust sterility assurance system is in place, consisting of:

- good product design;
- the company having knowledge and control of the microbiological condition of starting materials and process aids (e.g., gases and lubricants);
- good control of the contamination of the process of manufacture to avoid the ingress of microorganisms and their multiplication in the product. This is usually accomplished by cleaning and sanitation of product contact surfaces, prevention of aerial contamination by handling in cleanrooms or in isolators, use of process control time limits and, if applicable, filtration stages;

- systems for the prevention of mix-up between sterile and nonsterile product streams;
- maintenance of product integrity;
- a robust and consistent sterilization process;
- the totality of the quality system that contains the sterility assurance system (e.g., change control, training, written procedures, release checks, planned preventive maintenance, failure mode analysis, prevention of human error, validation, and calibration).

12.6   Sterile products

There are two main groups of sterile products, related to the way in which they are treated (or not) after being filled into the final container (be that a bag, vial, or syringe). The distinction is between products that can terminally sterilized in their final container and those that cannot due to the effect of the sterilization process upon the product. For example, some protein-based products cannot be subjected to heat. Products that cannot be subjected to terminal sterilization are aseptically filled and rely on the presterilization of the components and bulk product before being aseptically filled within a cleanroom. For these processes, there are different, and higher, levels of risk.

The regulatory bodies, such as the FDA and European regulators, favor terminal sterilization, and, in the development of new sterile dosage forms, the EU regulations demand that a decision tree is followed whereby the new dosage must be proven to be unable to withstand various defined processes of terminal sterilization before it is allowed to be manufactured aseptically. It is important that the organization has selected the appropriate method of sterile manufacturing and is aware of why that method is in place. The preparation of sterile products up to the filling and sterilization of the final product are broadly similar. The two types of sterile product are examined further below.

12.6.1   Terminal sterilization

Both the FDA guidance on aseptic filling (2004) and the European Pharmacopoeia (in Chapter 5.1.1) state that of the methods of sterile manufacture a process in which the product is sterilized in its final container (terminal sterilization) is the preferred method. This is not possible for all types of products and for this filtration through a bacteria-retentive filter and aseptic processing is used.

Terminal sterilization involves filling and sealing product containers under high-quality environmental conditions. This means that non-parenteral products that are to be terminally sterilized may be filled in an EU GMP Grade C/ISO 14644 class 8 area (for detail of cleanroom grades, see Chapter 16). With parenteral products these can be filled under the same conditions if the process or product does not pose a high-risk of microbial contamination. Examples of high-risk situations include slow-filling operations, the use of wide-necked containers or the exposure of filled containers to the environment for more than a few seconds before sealing. In these cases, products are filled in an aseptic area with at least an EU GMP Grade B/ISO 14644 class 7 environments or in an EU GMP Grade A/ISO 14644 class 5 zone with at least a Grade C/ISO class 8 background, prior to terminal sterilization.

Products are filled and sealed in this type of environment to minimize the microbiological content of the in-process product and to help ensure that the subsequent sterilization process is successful. It is accepted that the product, container, and closure will probably have low bioburden, but they are not sterile. The product in its final container is then subjected to a terminal sterilization process such as heat or irradiation. As terminally sterilized drug product, each product unit undergoes a single sterilization process in a sealed container. The assumption is that the bioburden within the product can be eliminated by the sterilization process selected [13].

Product formulation is undertaken at an EU GMP Grade C/ISO 14644 class 8 or an EU GMP Grade D/ISO 14644 class 9 environment. For some higher risk products a pre-filtration through a bacteria-retentive filter may be advisable in cases, particularly where there is a high bioburden. It is up to the pharmaceutical organization to define the level of risk and to justify this to an inspector.

12.6.2 Aseptic filling

Aseptic manufacturing is used in cases where the drug substance is instable when subjected to heat (thus, sterilization in the final container closure system is not possible) or where heat would cause packaging degradation. Aseptic filling is arguably the most difficult type of sterile operations. This is because the end product cannot be terminally sterilized and, therefore, there are far greater contamination risks during formulation and filling. With aseptic processing, there is always a degree of uncertainty, particularly because of the risk posed by personnel to the environment in which filling takes place.

In aseptic manufacture, the dosage form and the individual components of the containments system are sterilized separately, and then the whole presentation is brought together by methods that ensure that the existing sterility is not compromised. Sterility is normally achieved through sterile filtration of the bulk using a sterilizing grade filter (with a pore size of 0.2 μm or smaller) in sterile container closure systems and working in a clean area [14]. This is undertaken in an EU GMP Grade C/ISO 14644 class 8 cleanroom environment. The container and closure are also subject to sterilization methods separately. The sterilized bulk product is filled into the containers, stoppered and sealed under aseptic conditions (under EU GMP Grade A/ISO 14644 class 5 air) within an EU GMP Grade B/ISO 14644 class 7 cleanroom, unless filling is undertaken within a barrier system.

To assist with aseptic processing, engineering and manufacturing technology throughout all industries have evolved considerably. In the context of sterile and aseptic manufacture of pharmaceutical and medical devices, blow-fill-seal (BFS), prefilled syringe filling, restricted access barrier systems (RABS), and isolator technologies represent the main developments. Aseptic processes that exclude human intervention (such as robotics or barrier systems) are at a considerably lower risk than operations that consist of filling machines under unidirectional airflow devices where there is a need for periodic human intervention. With isolator systems the background environment for the cleanroom can be at EU GMP Grade C/ISO 14644 class 8, based on an appropriate risk assessment. There are additional risk considerations for isolators in that the decontamination procedures should be validated to ensure full exposure of all isolator surfaces to the chemical agent. Aseptic filling is the subject of Chapter 14.

12.6.3 Blow-fill-seal technology

BFS technology is a type of aseptic filling but one at a theoretical lower risk compared with conventional filling. BFS is an automated process where containers are formed, filled, and sealed in a continuous operation without human intervention. This is performed in an aseptic enclosed area inside a machine. The technology can be used to manufacture aseptically certain pharmaceutical liquid dosage forms.

BFS operations are undertaken under EU GMP Grade A/ISO 14644 class 5 conditions with the background environment at EU GMP Grade C/ISO 14644 class 8. Where BFS equipment is used for the production of products that are terminally sterilized, the operation can be carried out within an EU GMP Grade D/ISO 14644 class 9 background environments if appropriately risk assessed [15].

12.7   Sterilization

Although there is a wide variety of mechanisms and processes by which a pharmaceutical or medical device might be rendered free from microorganisms (i.e., sterile), they may be grouped into three main categories. These are [16]:

- *Physical removal*: the complete removal of all microorganisms to achieve a physical absence of microorganisms (such as filtration);
- *Physical alteration*: including physical destruction, disintegration of microorganisms. Altering, changing or deforming the physical cellular or biochemical architecture to destroy all physiological functionality;
- *Inactivation*: the permanent disruption of critical biochemical and physiological properties, potential and the microorganisms propensity (whether active or latent) to realize a clinical condition. Thus, ensuring impotency for generating an infection. For complete assurance of inactivation the microorganisms must, therefore, be essentially "killed" with no residual metabolic activity.

Sterilization is not the same as disinfection. Disinfection is a process that is designed to kill actively growing and vegetative microbial microorganisms to a certain level, and it does not, unless the disinfectant is classified as a sterilant, apply to bacterial endospores. Importantly, disinfection is not a substitute for sterilization [17]. Disinfection is the subject of Chapter 15.

From these important concepts, primary methods of sterilization consist of the following four main categories:

- high temperature/pressure sterilization, principally by dry heat or moist heat. Dry heat sterilization technology is less destructive to many materials than steam, which can be corrosive to metal objects and damaging to certain glass surfaces. However, the heating and cooling times of dry heat sterilizers often are lengthy. Dry heat at 160 °C (holding temperature for 1 h is required to kill the most resistant spores). A hot air oven is one of the most common methods used for dry heat sterilization.

Steam sterilization is performed under high pressure at temperatures that range from 121 to 140 °C, which is lower than temperatures required for dry heat sterilization and the sterilization times are often shorter. Steam (100 °C) is more effective than dry heat at the same temperature as: bacteria are more susceptible to moist heat, steam has more penetrating power, and steam has more sterilizing power as more heat is given up during condensation.

The classic method of sterilization by moist heat is autoclaving. With this method, sterilization is achieved by steam under pressure, to allow for steaming at temperatures higher than 100 °C. The theory runs that the temperature of boiling depends on the surrounding atmospheric pressure. A higher temperature of steaming is obtained by employing a higher pressure. When the autoclave is closed and made air-tight, and water starts boiling, the inside pressures increase and now the water boils above 100 °C. At 15 pounds per square inch (103 kPascal) of pressure, 121 °C temperature is obtained. This is maintained for 15 min for sterilization to kill spores. It works like a pressure cooker.

- Chemical sterilization: such as gassing using ethylene oxide. The active agent of the gas sterilization process can be ethylene oxide or another highly volatile substance. The sterilizing efficiency of a chemical such as ethylene oxide depends on the concentration of the gas, the humidity, the time of exposure, the temperature, and the nature of the load. In particular, it is necessary to ensure that the nature of the packaging is such that the gas exchange can take place.

The microbicidal activity of ethylene oxide gas is the result of alkylation of protein, deoxyribonucleic acid (DNA), and ribonucleic acid (RNA). Alkylation, or the replacement of a hydrogen atom with an alkyl group, within cells prevents normal cellular metabolism and replication.

An alternative gaseous process is hydrogen peroxide. Within the pharmaceutical industry, this gas is commonly used for the biodecontamination of isolators. Hydrogen peroxide inactivates microorganisms primarily by the combined use of the gas and the generation of free radicals (hydroxyl and hydroproxyl free radicals) during the plasma phase of the cycle.

- Filtration. Filtration concerns the rendering of a liquid or gas as sterile by the physical removal of microorganisms. In order to remove bacteria, the membrane pore size (primarily 0.22 μm) must be smaller than the bacteria and uniform throughout. Filtration processes can be complicated, and the validation is dependent upon the type of material being filtered and its physical properties.
- Radiation sterilization: such as gamma radiation and electron beam. These two radiation processes have similarities and differences and are applicable to different situations and are suited for different materials, in relation to speed of processing, degree of penetration and validation requirements.

There are two types of radiation used for sterilization: ionizing radiation and nonionizing radiation. Ionizing radiation is the use of short wavelength, high-intensity radiation to destroy microorganisms. This radiation can come in the form of gamma- or X-rays that react with DNA resulting in a damaged cell. Nonionizing radiation uses longer wavelength and lower energy. As a result, nonionizing radiation loses the ability to penetrate substances and can only be used for sterilizing surfaces. The most common form of nonionizing radiation is ultraviolet (UV) light.

The process of selecting the appropriate sterilization method is driven by regulatory, economic, and scientific requirements. In considering the scientific and practical aspects, factors to consider include:

(a) whether the product and its packaging fit into an existing sterilization technology. Ideally, a product will tolerate several different types of technology;
(b) the logistics of transporting the product to and from the site of sterilization. This is obviously easier when the sterilization takes place in-house compared with transporting the product to a contract facility;
(c) validation requirements to verify that the sterilization cycle is effective.

All forms of sterilization have negative effects to a wide variety of packaging materials (and sometimes on the item or product itself). These effects can vary from material to material and between the different packaging components. Sterilization can affect polymers, seal strength, label and box adhesion, corrugated and paperboard strength, and material color. The selection of the sterilization method is, therefore, of considerable importance.

The validation of most, but not all, sterilization processes requires a biological control to supplement physical measurement. This is particularly important for sterilization by heat. For this, biological indicators (spore populations) of bacteria of a known resistance are used. With biological indicators, one common way of quantifying microbial death to a sterilization process is the D value. The D value is the time or dose required to reduce the microbial population by one log (or 90%, e.g., the time or dose required to reduce a population of 1000–100 cells) [18]. Biological indicators are discussed in Chapter 13.

12.8   Factors affecting sterilization effectiveness

There are a number of factors that affect the success or otherwise of a sterilization process. These are outlined below.

12.8.1   Number and location of microorganisms

All other conditions remaining constant, the larger the number of microorganisms then the time, a sterilization process is required to run for, becomes longer, in order to destroy all of microorganisms present. Reducing the number of microorganisms that must be inactivated through meticulous cleaning and disinfection, or by assembling components within classified cleanrooms, increases the margin of safety when the sterilization process is applied.

In terms of the location of microorganisms, research has shown that aggregated or clumped microbial cells are more difficult to inactivate than monodispersed cells. Microorganisms may also be protected from poor penetrating sterilization methods by the production of thick masses of cells and extracellular materials, or biofilms [19]. It has also been shown that products that have crevices, joints, and channels are more difficult to sterilize than are flat-surface equipment because penetration of the sterilizing agent to all parts of the equipment is more difficult.

12.8.2 Innate resistance of microorganisms

Microorganisms vary greatly in their resistance to sterilization processes. Intrinsic resistance mechanisms in microorganisms vary. For example, spores are generally the most resistant to sterilization processes because the spore coat and cortex act as a barrier. Implicit in all sterilization strategies is the consideration that the most resistant microbial sub-population controls the sterilization time. That is, to destroy the most resistant types of microorganisms (bacterial spores), the user needs to employ exposure times and a concentration or dose needed to achieve complete destruction [20].

12.8.3 Physical and chemical factors

Several physical and chemical factors also influence sterilization processes, especially temperature and relative humidity. For example, relative humidity is the single most important factor influencing the activity of gaseous sterilants, such as ethylene oxide, chlorine dioxide, and formaldehyde [21]. Whereas achieving a certain temperature is critical for the operation of an autoclave.

12.8.4 Organic and inorganic matter

Organic matter, such as serum or blood, can interfere with the antimicrobial activity of sterilization processes by interfering with the chemical reaction between the certain sterilants and the organic matter resulting in less of the active sterilant being available for attacking microorganisms. The effects of inorganic contaminants on the sterilization process can afford protection to microorganisms thereby limiting the potential effectiveness of the sterilization process.

12.8.5 Duration of exposure

Items must be exposed to the sterilization process for an appropriate minimum time. Most sterilization processes have minimum cycle times, established during validation runs.

12.8.6 Storage

All sterile items should be stored in an area and manner whereby the packs or containers will be protected from dust, dirt, moisture, animals, and insects. The shelf life of sterilization depends on the following factors:

- quality of the wrapper or container;
- number of times a package is handled before use;

- number of people who have handled the package;
- whether the package is stored on open or closed shelves;
- condition of storage area (e.g., humidity and cleanliness);
- use of plastic dust covers and method of sealing.

12.9   Good manufacturing practice

Sterile processing is one of the most regulated areas within healthcare. The reason is due to the potential risk to patients, either directly from the developed drug or through an associated reagent or medical device. If the medicine is contaminated with microorganisms, then the patient may become ill or even die [22].

Any manufacturer of a sterile medicinal product or who operates a sterilization process that links into the production of a sterile product will be subject to a regulatory inspection from their national agency and from overseas agencies if the product is intended for distribution into territories that fall under the auspice of a particular agency. Most regulators adopt a risk-based approach to regulations, guidelines, and inspections. Consequently, risk assessment should be firmly built into the pharmaceutical organization's quality system. Risk management is fundamentally about understanding what is most important for the control of product quality and then focusing resources on managing and controlling these things to ensure that risks are reduced and contained. Before risks can be managed, or controlled, they need to be assessed.

Two important points to remember for any risk assessment are that, first, there is no such thing as "zero risk," and therefore, a decision is required as to what is "acceptable risk." Second, risk assessment is not an exact science—different people will have a different perspective on the same hazard [23].

Regulations relating to pharmaceuticals and microbiological aspects are addressed more widely in Chapter 3.

12.10   Risk assessment

When considering any types of sterile manufacturing, the essential risk must never be forgotten: that the objective is to avoid the contamination of the product by microorganisms or microbial by-products (such as endotoxins). It is also important to focus on the most common sources of contamination. These are [24]:

- *Air*: air is not a natural environment for microbial growth (it is too dry and absent of nutrients), but microorganisms such as *Bacillus*, *Clostridium*, *Staphylococcus*, *Penicillin*, and *Aspergillus* can survive. To guard against this, products and sterile components must be protected with filtered air supplied at sufficient volume;
- *Facilities*: inadequately sanitized facilities pose a contamination risk. Furthermore, poorly maintained buildings also present a risk such as potential fungal contamination from damp or inadequate seals. The design of buildings and the disinfection regimes are thus of importance;

- *Water*: the presence of water in cleanrooms should be avoided. Water is both a growth source and a vector for contamination;
- *Incoming materials*: incoming materials, either as raw materials (which will contain a level of bioburden) or as packaged materials, present a contamination risk if they are not properly controlled. Paper and cardboard sources in particular present a potential risk;
- *People*: people are the primary source of contamination within cleanrooms. People generate millions of particles every hour from activities of breathing, talking, and body movements, where particles are shed from hair, skin, and spittle. Many of these particles will be carrying microorganisms.

These factors should be borne in mind when designing different sterilization processes.

12.11 Conclusion

This chapter has provided an introduction to sterility and to sterilization. The chapter has outlined pharmaceutical microbiology in relation to the ways by which microorganisms can survive within processing environments and thereby present a risk to sterilization or to aseptic filling. The chapter has explained further that sterility is an absolute term, but equally one that is difficult to prove, and thus, it can only be understood in terms of risk and probability. For terminally sterilized products and sterilization processes the sterility assurance concept is useful. This concept cannot, however, be applied to aseptic filling, and instead there is a strong reliance upon environmental controls.

References

[1] Akers MJ. Introduction, scope, and history of sterile products. In: Akers MJ, editor. Sterile drug products: formulation, packaging, manufacturing and quality. London: Informa Healthcare; 2010. p. 1–11.

[2] Sandle T, Saghee MR. The essentials of pharmaceutical microbiology. In: Saghee MR, Sandle T, Tidswell EC, editors. Microbiology and sterility assurance in pharmaceuticals and medical devices. New Delhi: Business Horizons; 2011. p. 1–30.

[3] Favero MS. Sterility assurance: concepts for patient safety. In: Rutala WA, editor. Disinfection, sterilization and antisepsis: principles and practices in healthcare facilities. Washington, DC: Association for Professional in Infection Control and Epidemiology; 2001. p. 110–9.

[4] International Organization for Standardization. Sterilization of health care products—vocabulary. ISO/TS 11139:2006. Geneva: International Organization for Standardization; 2006.

[5] Sandle T. Sterility, sterilisation and sterility assurance for pharmaceuticals: technology, validation and current regulations. Cambridge, UK: Woodhead Publishing; 2013. p. 1–20.

[6] Berube R, Oxborrow GS, Gaustad JW. Sterility testing: validation of sterilization processes and sporicide testing. In: Block S, editor. Disinfection, sterilization and preservation. 5th ed. Philadelphia, PA: Lippincott Williams & Wilkins; 2001. p. 1361.

[7]   Baird RM. Sterility assurance: concepts, methods and problems. In: Russell AD, Hugo WB, Ayliffe GAJ, editors. Principles and practice of disinfection, preservation and sterilization. 3rd ed. Oxford: Blackwell Science; 1999. p. 787–99.

[8]   Stumbo CR. Thermobacteriology in food processing. 2nd ed. Orlando, FL: Academic Press; 1973. p. 130.

[9]   Mosley GA. Microbial lethality: when it is log-linear and when it is not! Biomed Instrum Technol 2003;37(6):451–4.

[10] Tidswell EC, Khorzad A, Sadowski M. Novel and emerging sterilization technologies. Eur Pharm Rev 2009;5:11–21.

[11] Sandle T. Sterility test requirements for biological products. Pharm Microbiol Forum Newsl 2011;17(8):5–14.

[12] PIC/S. Recommendation on guidance for parametric release. Brussels: Pharmaceutical Inspection Convention and Pharmaceutical Inspection Co-operation Scheme 2007. At http://www.picscheme.org/pdf/24_pi-005-3-parametric-release.pdf.

[13] Agallocco J. Process selection for sterile products. In: Saghee MR, Sandle T, Tidswell EC, editors. Microbiology and sterility assurance in pharmaceuticals and medical devices. New Delhi: Business Horizons; 2011. p. 603–14.

[14] Meltzer TH. Filtration in the pharmaceutical industry. New York: Marcel Dekker; 1987. p. 90–5.

[15] Bradley A, Probert SC, Sinclair CS, Tallentire A. Airborne microbial challenges of blow/ fill/seal equipment: a case study. J Parenter Sci Technol 1991;45:187–92.

[16] Tidswell E. Sterility. In: Saghee MR, Sandle T, Tidswell EC, editors. Microbiology and sterility assurance in pharmaceuticals and medical devices. New Delhi: Business Horizons; 2011. p. 589–602.

[17] Sandle T. Cleaning and disinfection. In: Sandle T, editor. The CDC handbook: a guide to cleaning and disinfecting clean rooms. Surrey, UK: Grosvenor House Publishing; 2012. p. 1–31.

[18] Akers MJ, Anderson NR. Sterilization validation. In: Nash RA, Wachter AH, editors. Pharmaceutical process validation. 3rd ed. New York: Marcel Dekker; 2003. p. 83–157.

[19] LeChevallier MW, Cawthon CD, Lee RG. Inactivation of biofilm bacteria. Appl Environ Microbiol 1988;54:2492–9.

[20] Russell AD. Principles of antimicrobial activity and resistance. In: Block SS, editor. Disinfection, sterilization, and preservation. Philadelphia, PA: Lippincott Williams & Wilkins; 2001. p. 31–55.

[21] Rutala WA. Selection and use of disinfectants in healthcare. In: Mayhall CG, editor. Hospital epidemiology and infection control. Philadelphia, PA: Lippincott Williams & Wilkins; 1999. p. 1161–87.

[22] Rutala WA. Antisepsis, disinfection and sterilization in the hospital and related institutions. In: Balows A, editor. Manual of clinical microbiology. Washington, DC: ASM; 1995. p. 227–45.

[23] Sandle T. Risk management in pharmaceutical microbiology. In: Saghee MR, Sandle T, Tidswell EC, editors. Microbiology and sterility assurance in pharmaceuticals and medical devices. New Delhi: Business Horizons; 2011. p. 553–88.

[24] Reinmüller B. People as a contamination source—clothing systems. In: Anon., editors. Dispersion and risk assessment of airborne contaminants in pharmaceutical clean rooms. Stockholm: Royal Institute of Technology, Building Services Engineering; 2001. p. 54–77 [Bulletin No. 56].

Biological indicators: Measuring sterilization

13

13.1 Introduction

Biological indicators include preparations of selected microorganisms with high resistance towards specific sterilization methods [1]. Since sterilization is only a probability of there being an absence of microorganisms (where true sterility can only be demonstrated through infinite exposure), then to seek greater assurance for sterilization can be measured through physical (thermometric) data and from biological indicators to provide confidence that a sterilization process has been successful. There is a recurrent debate as to whether physical data or biological data is the most important when examining sterilization processes, as well as considering if both measures are needed. Although sterilization can be measured thermometrically using thermocouples, it is increasingly common for the validation and routine re-qualification to use biological indicators. Theoretically, the reason for this is dry heat in comparison with moist heat (saturated steam). A thermocouple might be reading the correct temperature, however, the local environment may consist of a "dry pocket." A good practice is a combination of a biological indicator and a thermocouple to ensure saturated steam exists in the region where the thermocouple was placed [2].

Biological indicators are "standardized" preparations of certain microorganisms with known characteristics. The microorganisms used to prepare biological indicators are those capable of forming endospores, and the microorganism is used in the "spore state." A biological indicator is prepared by depositing bacterial spores, from a spore crop, are deposited onto a carrier, such as filter paper. The carrier may be wrapped in a suitable primary package. In preparing a biological indicator, the object is to use a microorganism of a known population, purity, and resistance characteristic [3].

The resistance of microorganisms to sterilization treatments differs according to a range of factors. Some of these factors may be intrinsic to the microorganism itself, others may be a combination of intrinsic and extrinsic factors such as the chemistry of the carrier. The principle behind the use of biological indicators is that if bacterial spores are destroyed, then it can be assumed that any contaminating microorganisms in the sterilization load would also have been killed, as these microorganisms will have lower resistance than any spores that might be present (and such environmental microorganisms will have been present in far lower numbers).

13.2 Origins

The term "biological indicator" has wide usage outside of the pharmaceutical industry. The term biological indicator (or "bioindicator") is also applied generally to the application of plants or animals to various conditions where the reaction of the biological

Pharmaceutical Microbiology. http://dx.doi.org/10.1016/B978-0-08-100022-9.00013-X

material is examination. In one sense, the use of a canary in a cage by a miner to detect pockets of natural gas was arguably one of the first biological indicators. To assess sterility, in similar presentation and form, biological indicators are commonly used in both food and pharmaceutical industries [4].

The original format of biological indicators was inoculated paper strips inside envelopes, which were transferred to sterile culture medium following processing and incubated for 7 days. Sterilization failure was measured by turbidity of the growth medium. From this starting point, some organizations elect to incubate biological indicators for up to 14 days, and second-generation biological indicators are commonly used (such as self-contained systems that comprise the microorganism and growth medium required for recovery in a primary pack ready for use. Microbial growth is indicated by a change in pH (with a color indicator), which measures the production of acid metabolites in the growth medium by outgrowing spores and replicating microbial cells).

13.3 Types of biological indicators

Different biological indicators are used for different sterilization processes. Biological indicators are designed for use with:

- ethylene oxide gas;
- hydrogen peroxide vapor;
- dry heat;
- steam;
- radiation.

With each of these [5]:

- ethylene oxide gas is used to kill bacteria, mould, and fungi in medical supplies such as bandages;
- dry-heat sterilization uses an oven to raise the temperature of items that are wrapped in foil or fabric;
- steam sterilization uses an autoclave, a self-locking machine that sterilizes its contents with steam under pressure;
- irradiation is used to sterilize materials that may be damaged by moist heat, such as plastics.

The microorganism used to prepare the biological indicator will vary depending upon the means of sterilization that requires testing. Microorganisms are selected depending upon how resistant they are to the chosen method of sterilization. Different microorganisms are more resistant than others to different types of sterilization. With steam sterilization, for example, spore-bearing microorganisms are more resistant than nonspore-bearing microorganisms. A microorganism such as *Staphylococcus* (commonly carried on human skin) would have a typical D value at 121 °C for a 15-min autoclave cycle of only 15 s, whereas an endospore-forming thermophilic *Bacillus* would have one of at least 1.5 min. For steam sterilization, *Geobacillus stearothermophilus* is most commonly used (as required by the pharmacopoeia). This microorganism is used due to its theoretical resistance to particular types of sterilization, including heat.

Biological indicators are available in many different forms. Examples include strips (the classic "spore strip"), discs, suspensions, test tubes, and ampoules. With these:

- spore strips are biological indicators that are packaged in a pouch made of glassine, a paper that is resistant to moisture and air at ambient temperatures and pressures;
- spore discs are usually made of borosilicate paper or stainless steel. Spore suspensions are diluted aliquots that are derived from a primary batch of spores. Other spore suspensions that are inoculated directly onto surfaces, such as rubber closures;
- test tubes that are available in a variety of sizes and are usually made of expansion-resistant glass. Ampoules are small, self-contained vials that are hermetically sealed with a flame. They have a score mark around the neck so that the sealed top can be snapped off by hand. Typically, ampoules are used to contain hypodermic injection solutions.

13.4 Characteristics of biological indicators

The most important characteristic of a biological indicator is that sporulation must readily occur on a defined medium and, if there are any survivors, spore germination will occur [6]. If there are any survivors, then it is important that the survivors form easily countable colonies. Without possessing these characteristics, then the biological indicator is of little value. As this is a critical parameter, it is recommended that a positive control be run alongside each test set of biological indicators.

All biological indicators must come with a certificate of conformity. The certificate should indicate the population, D value, and purity of the microorganism. Due to the variability in the preparation of biological indicators, some users elect to have biological indicators verified (this would be the case with, for example, spores inoculated onto a paper carrier to create a spore strip). Biological indicators that the user prepares (such as inoculating a spore suspension onto a rubber closure) must always be verified, as there is no other comparative data available.

Each of the key parameters for biological indicators is examined below.

13.4.1 Purity

Although some biological indicators may contain other microorganisms, when subjected to an appropriate challenge only the intended microorganism should be recovered. For example, using a heat shock test with biological indicators for steam sterilization, the only thermophilic microorganism detected should be *G. stearothermophilus*. Biological indicators must be verified for purity by at least a phenotypic identification of the microorganism.

13.4.2 Population

Biological indicators must have a minimum population as defined by the pharmacopoeias. The target population for biological indicators is ordinarily greater than 10^6. The reason this population is used is because it is generally accepted that "devices purporting to be sterile," such as an autoclave, are designed to achieve a 10^{-6} microbial

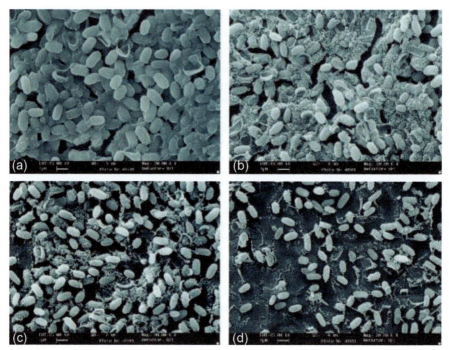

Figure 13.1 *Geobacillus stearothermophilus* spores on a carrier. (a) Untreated strip; (b) spore strip subject to 30 s of heat at 121°C; (c) application of 60 s of heat; and (d) application of 120 s of heat.
Image courtesy of pharmig.

survival probability (i.e., there is less than one chance in a million that a microorganism would survive the sterilization process).

A population verification, ISO 11138 total viable spore count, is normally performed [7]. The acceptance criteria state that the results should be no less than 50% or more than 300% of the labeled certified population (Figure 13.1).

13.4.3 D *value*

Arguably the most important characteristic of biological indicators is the level of resistance. This is defined by the decimal reduction value (or *D* value) [8].

The *D* value is the time taken to reduce the population of a known microorganism by 1 log (or 90% of the population). Thus, after an organism is reduced by 1*D*, only 10% of the original microbial population remains (i.e., the population number has been reduced by one decimal place in the counting scheme). When referring to *D* values, it is normal to give the temperature as a subscript to the *D*. For example, a hypothetical organism is reduced by 90% after exposure to temperatures of 121 °C for 1.5 min. Thus, the *D* value would be written as $D_{121°C} = 1.5$ min. *D* values will vary according to the resistance of the microorganism and the population challenge. Generally, the longer the exposure time and the more resistant the microorganism, the higher the D value.

Once a D value has been established, many sterilization cycles have "overkill" built in. This is either simply doubling the cycle time (or sterilization dose), or it is taken from a mathematically calculated SAL. Typically the SAL is developed to give a sterilization cycle designed to achieve a 12-log reduction of the challenge population [9]. For instance, many autoclaves are operated with an overkill cycle, where the temperature or time of the cycle is increased with the aim of achieving a greater theoretical kill of the challenge biological indicator. Anderson argues that the overkill cycle compensates for deviations in time or calibration faults with thermocouples, variability in packaging, or chamber leaks [10].

The acceptance criteria for the D value are defined by the US Pharmacopeia (USP), which states:

The requirements of the test are met if the determined D-value is within 20% of the labelled D-value for the selected sterilizing temperature and if the confidence limits of the estimate are within 10% of the determined D-value.

In order to verify the D Value, the USP and ISO 11138-14 allow for the use of three methods. These are:

- the most probable number method by direct enumeration;
- a fraction negative method (such as Spearman/Karber);
- assessing the D value accuracy by using the USP survive/kill calculated cycles.

Regardless of which of the three methods is used, the piece of equipment that will be needed to calculate the D value is a resistometer. A resistometer, also known as a BIER Vessel (Biological Indicator-Evaluator Resistometer), is an item of test equipment that can very quickly and accurately deliver and control very precise sterilization process parameters. With a Steam BIER Vessel, the equipment must be capable of reaching the target temperature set point within 10s or less from the time "steam charge" occurs. Additionally, it must maintain that set temperature to within $\pm 0.5\,°C$ and then at cycle end, the postvacuum time to reach atmospheric pressure must be within 10s or less [11].

The most common method deployed to calculate D values is a fraction negative method. For this method, multiple groups of biological indicators (typically 10 or 20) are exposed to varying cycle exposure times. The examination is for partial kill (looking for that fraction which is negative). This is normally running one exposure designed for all test biological indicators to survive, one exposure designed for all test biological indicators to be killed, and several exposures in between, set at equidistant time intervals.

For example, to verify the resistance of a particular biological indicator in a steam vessel at 121 °C using the limited Spearman–Karber fraction negative method, about 20 biological indicators would be exposed per group to various exposure times at 121 °C. After each exposure, each group of biological indicators would be aseptically transferred to a growth medium and incubated at the appropriate temperature.

D values vary with different carriers, even where the same spore crop is used. Thus, the same spore crop used to inoculate a paper strip and a rubber closure will give a different D value (and there is a likely probability that the rubber closure will give a higher D value). This variation explains why, for instance, the D value for a self-contained biological indicator in a glass ampoule has a higher D value than spores inoculated onto a cotton thread.

A similar phenomenon occurs with fluids. Spores suspended in water will have a lower D value than spores suspended in a saline solution.

13.4.4 Z value

A Z value is defined as the number of degrees Celsius required to change a D value by one factor of 10. In the practical sense, it is a measure of how susceptible a spore population is to changes in temperature. For example, if the Z value of a population is 10 °C, then increasing the sterilization temperature 10 °C will result in a log reduction of the D value.

To work out a Z value, at least three D value/temperature pairs are required. Z values can be estimated graphically (using line of best fit) or calculated mathematically. Z values are useful for calculating F values (in conjunction with D values), especially to show the relationship between lethalities.

Other factors need to be considered when using biological indicators. These include the shelf life, strip size, and package size of the biological indicator.

13.4.5 Assessing results

The way in which results from biological indicator testing are handled is to plot a graph (the survivor curve) with the logarithm of the number of viable microorganisms on the ordinate and the time of exposure on the abscissa. This is called a semilogarithmic plot. The general form of inactivation of microbial populations when plotted semilogarithmically is linear. The terms used to describe this are logarithmic or exponential inactivation or death.

Although most sterilization devices are well designed, there are a number of occasions where biological indicators can fail. For example, a positive test result from a biological indicator can result from a variety of causes, such as inadequate steam quality, insufficient exposure time or temperature, poor loading practices, or product failure or operator failure [12].

13.5 Testing issues

When setting up a biological indicator study, there are a number of issues that need to be considered in advance of undertaking the validation. These include:

- the number of biological indicators required should be assessed upfront;
- the locations for the biological indicators should be considered in advance;

- the location where biological indicators are placed in relation to thermocouples should be considered, especially if this might affect air removal, steam penetration, condensate collection, or air leakage.

Furthermore, personnel carrying out biological indicator tests must be trained, especially in demonstrating that they can reproduce counts within a range of variability [13].

13.6 Areas of concern and testing errors

As with any biological test, there are aspects of biological indicator testing which can cause testing difficulties [14]. Some of these issues are next examined.

(a) The bioburden of the product being sterilized can affect the results of the study, such as leading to an increase in the D value or promoting survival of spores through a clumping effect by one microorganism covering another. Therefore, the following should be considered:
   - total numbers of organisms present, as the item to be sterilized, just prior to sterilization must be known;
   - types of organisms present;
   - number of resistant spore formers present;
   - resistance of this bioburden;
   - sampling frequency and statistical analysis.

(b) Variability between different lots of biological indicators.

Each lot of biological indicators will vary slightly in its population, resistance, and kill time. This variability can arise from heterogeneity within a spore population, which can be caused by genotypic and phenotypic variations within the spore crop. This is one of the reasons why the USP recommends that supplier audits take the place of biological indicator manufacturers. In addition, it is good practice to audit any contract test laboratories that may undertake biological indicator testing.

(c) Shipping conditions.

Biological indicators may be affected by the transport from the manufacturer. Any available transport and stability data from manufacturer should be reviewed.

(d) Storage conditions.

Most biological indicators will have prescribed storage conditions. These may be strictly defined, or "controlled temperatures" will be referred to. Controlled room temperature is defined in the USP as:

> A temperature maintained thermostatically that encompasses the usual and customary working environment of 20 to 25 °C (68 to 77 °F); that results in a mean kinetic temperature calculated to be not more than 25 °C; and that allows for excursions between 15 and 30 °C (59 to 86 °F) that are experienced in pharmacies, hospitals and warehouses...

Humidity, if it is not defined by the manufacturer, is typically 20–70% relative humidity.

Storage conditions and times should be assessed by a stability trial. This is of great importance as, theoretically, the *D* value of a biological indicator will decrease over time.

(e) Delay in transferring the biological indicator to storage medium.

Theoretically, the ability to recover spores, especially those that are sub-lethally damaged, may be affected by the time taken to transfer a biological indicator that has undergone steam sterilization to the required culture medium. For this purpose, the USP states in the *Guide to General Chapters Microbiological Tests <55> Biological Indicators*, that:

> *...after completion of the sterilizing procedure... and within a noted time not greater than 4 hours, aseptically remove and add each strip to 10 to 30 ml of Soybean Casein Digest medium...*

(f) Test method used by contract test laboratory to determine the *D* value.

Variation can arise when biological indicators are evaluated by contract manufacturers for population and *D* value. Variables can include techniques, utensils, and equipment. The main source of variation is if the contract test laboratory uses a different technique for *D* value determination from the manufacturer [15]. A related variation can arise from the culture medium, and incubation conditions for different brands and different lots of culture media may not have the same degree of ability to promote growth of injured spores [16].

(g) Preparation of biological indicators

Variation can occur with the preparation of biological indicators. This is of particular concern when users prepare their own biological indicators, such as inoculating spores onto stoppers. Areas of concern here include:

- how spores are put onto carriers;
- places where the inoculation is too thick (and irregular clumps occur);
- how often the spore suspension is re-suspended;
- pipetting technique;
- drying times;
- the fluid in which the spore suspension is held (typically water or ethanol);
- problems from media residues;
- excessive damage to the surface.

13.7 Summary

This chapter has examined some of the key characteristics of biological indicators. Biological indicators are of great importance in assessing sterilization in the pharmaceutical industry. Thermometric data provides abundant information as to what might theoretically happen; however, it is only through biological material that the question "what if my material to be sterilized has a high bioburden?" can be answered.

The emphasis of the chapter has been upon some of the factors that might cause variation and testing problems. An element of variation will always be present when biological material is used; however, attempts should be made to reduce this variation to a minimal level.

References

[1] Sandle T. Biological indicators. In: Handlon G, Sandle T, editors. Industrial pharmaceutical microbiology: standards & controls. Passfield, UK: Euromed Communications; 2014. p. 16.1–16.26.

[2] Agalloco JP, Akers JE, Madsen RE. Moist heat sterilization—myths and realities. PDA J Pharm Sci Technol 1998;52(6):346–50.

[3] Sandle T. Sterility, sterilisation and sterility assurance for pharmaceuticals: technology, validation and current regulations. Cambridge, UK: Woodhead Publishing; 2013, p. 263–78.

[4] Tidswell E. Sterility. In: Saghee MR, Sandle T, Tidswell EC, editors. Microbiology and sterility assurance in pharmaceuticals and medical devices. New Delhi: Business Horizons; 2010. p. 569–614.

[5] Joslyn LJ. Sterilization by heat. In: Block SS, editor. Disinfection, sterilization, and preservation. 4th ed. Philadelphia, PA: Lea and Febiger; 1991. p. 495–526.

[6] Foster SJ, Johnstone K. The trigger mechanism of bacterial spore germination. In: Smith I, Slepecky RA, Setlow P, editors. Regulation of prokaryotic development. New York: Plenum Publishing Corporation; 1989. p. 89–108.

[7] ISO 11138-3:2006. Sterilisation of health care products—biological indicators—part 3: biological indicators for moist heat sterilisation processes. Geneva: International Standards Organization; 2006.

[8] Pflug IJ, Odlaug TE. Biological indicators in the pharmaceutical and medical device industry. J Parenter Sci Technol 1986;40:242–8.

[9] Sandle T. Biological indicators for steam sterilisation. Pharmig News 2009;34:10–4.

[10] Anderson J. Defining and presenting overkill cycle validation. Pharm Technol Eur 2013;25(11):26–7.

[11] Mosley GA, Gillis J. Operating precision of steam BIER vessels and the interactive effects of varying Z values on the reproducibility of listed D values. PDA J Pharm Sci Technol 2002;55(6):318–31.

[12] Sandle T. Biological indicators for steam sterilization: failure investigations. Pharm Microbiol Forum Newsl 2011;17(5):2–8.

[13] Moldenhaurer J. Biological indicator performance standards and control. In: Jimmenez L, editor. Microbiological contamination control in the pharmaceutical industry. New York: Marcel-Dekker Inc.; 2004. p. 133–45.

[14] Shintani H, Akers JE. On the cause of performance variation of biological indicators used for sterility assurance. PDA J Pharm Sci Technol 2000;54(4):332–42.

[15] Halls NA. Resistance "creep" of biological indicators. In: Morrissey RF, Kowalski JB, editors. Sterilization of medical products. Champlain, NY: Polysciences Publications, Inc.; 1998. p. 313–23.

[16] Penna TCV, Machoshvile IA, Taqueda MES, Ishii M. The effect of media composition on the thermal resistance of *Bacillus stearothermophilus*. PDA J Pharm Sci Technol 2000;54(5):398–412.

Antibiotics and preservatives

14

14.1 Introduction

An antimicrobial agent is a chemical that either kills or inhibits the growth of microorganisms. Each class of antimicrobial agents has a unique mode of action. Antimicrobial agents that inhibit microbes, such as tetracycline, have the suffix "static" added to their root (e.g., "bacteriostatic"); whereas agents that kill microbes, such as fluoroquinolones, have the suffix "cidal" added to their root (e.g., "bactericidal"). Antimicrobial agents include antibacterial, antiviral, antifungal, and antiparasitic agents.

The first antibiotic was serendipitously "discovered" in 1928 by Alexander Fleming when *Penicillium novatum* spores were observed to inhibit the growth of *Staphylococcus* on agar. The first wave of antibiotics was derived from microorganisms; latterly, following advances in medicinal chemistry, most modern antibacterials are semisynthetic modifications of various natural compounds [1]. These often differ from their parent compound in their antimicrobial activity or their pharmacological properties.

Antibiotic sensitivity is the susceptibility of bacteria to antibiotics. Antibiotic susceptibility testing (AST) is undertaken to determine which antibiotic will be most successful in treating a bacterial infection *in vivo*. Although it is more commonly an aspect of clinical microbiology, the performance of antimicrobial susceptibility testing of significant bacterial isolates plays an important part of the pharmaceutical development of new antimicrobials. The objective of this testing is to detect possible drug resistance in common pathogens and to assure susceptibility to drugs of choice for particular infections. The most widely used testing methods include broth microdilution or alternative rapid automated instrument methods that use commercially marketed materials and devices. Manual methods include the disc diffusion and gradient diffusion methods. Each method has strengths and weaknesses, although, in general, testing methods provide accurate detection of common antimicrobial resistance mechanisms. That said, newer or emerging mechanisms of resistance require constant vigilance regarding the ability of each test method to detect resistance accurately.

A preservative is a natural or synthetic chemical that is added to products such as foods, cosmetics, or pharmaceuticals to prevent spoilage [2]. A preservative is added to pharmaceuticals to prevent decomposition by microbial growth or by undesirable chemical changes. Antimicrobial preservatives are substances added to products to protect them from microbiological growth or from microorganisms that are introduced inadvertently during or subsequent to the manufacturing process.

Upon determination that a product has been properly neutralized and has very low levels of contamination (<10 colony-forming units (CFU)/g), microbiologists can conduct a preservative efficacy test (PET) (alternatively termed an antimicrobial efficacy test). With this test, the product is challenged within individual containers, separately, using one of the five required microorganisms. These organisms, as dictated by

Pharmaceutical Microbiology. http://dx.doi.org/10.1016/B978-0-08-100022-9.00014-1

compendia, are: *Escherichia coli, Staphylococcus aureus, Pseudomonas aeruginosa, Candida albicans,* and *Aspergillus brasiliensis.* The PET assay measures the reduction of a high inoculum (10^6–10^7 CFU/g) in the presence of a product containing preservative over 28 days. The method tests the ability of the preservative to prevent a re-bound of an organism (re-growth).

14.2   Antibiotic susceptibility testing

14.2.1   Antimicrobials

Resistance to antibiotics can either be naturally occurring for a particular organism and drug combination or acquired resistance can occur. This is where over-use (misuse) of antimicrobials results in a population being exposed to an environment in which organisms that have genes conferring resistance (either spontaneously mutated or through deoxyribonucleic acid (DNA) transfer from other resistant cells) have been able to flourish and spread.

Antimicrobial agents are classified by their specific modes of action against bacterial cells. These agents may interfere with cell wall synthesis, inhibit protein synthesis, interfere with nucleic acid synthesis, or inhibit a metabolic pathway. Mechanisms include:

- interference with cell wall synthesis (such as activity of beta-lactams);
- interference with the cytoplasmic membrane;
- interference with protein synthesis by binding to the 30S ribosomal subunit;
- inhibition of protein synthesis by binding to the 50S ribosomal subunit;
- inhibition of protein synthesis by inhibition of the 70S initiation complex;
- interference with nucleic acid synthesis;
- inhibition of the metabolic pathway for folic acid synthesis.

There are a number of ways by which microorganisms are resistant to antimicrobial agents. These include:

- the bacteria produce enzymes that either destroy the antimicrobial agent before it reaches its target or modify the drug so that it no longer is recognized by the target;
- the cell wall becomes impermeable to the antimicrobial agent;
- the target site is altered by mutation so that it no longer binds the antimicrobial agent;
- the bacteria possess an efflux pump that expels the antimicrobial agent from the cell before it can reach its target;
- specific metabolic pathways in the bacteria are genetically altered so that the antimicrobial agent cannot exert an effect.

With the above, in some species antimicrobial resistance is an intrinsic or innate property using one of the aforementioned mechanisms. Bacteria also can acquire resistance to antimicrobial agents by genetic events such as mutation, conjugation, transformation, transduction, and transposition.

These variable aspects of resistance make antimicrobial susceptibility testing of importance.

14.2.2 Antimicrobial susceptibility test concepts

Identification of a microorganism in the clinical setting normally goes hand in hand with the AST test. This is because knowing what microorganism has isolated together with knowledge of the isolation site, will give an indication of what type of antibiotics should be considered. The sensitivity of an isolate to a particular antibiotic is measured by establishing the minimum inhibitory concentration (MIC) or breakpoint. MIC is the lowest concentration of antibiotic at which an isolate cannot produce visible growth after overnight incubation.

MICs can be determined by agar or broth dilution techniques by following the reference standards established by various authorities. These include the Clinical and Laboratory Standards Institute (CLSI, USA), British Society for Antimicrobial Chemotherapy (BSAC, UK), AFFSAPS (France), Deutsches Institut für Normung e.V. (DIN, Germany), and ISC/World Health Organization (WHO).

With each method, one of the most important steps in the testing process is preparing the inoculum of the test microorganism. This involves selecting appropriate colonies for testing, suspending them in broth, and standardizing the suspension.

14.2.3 Broth dilution method

The broth dilution method depends upon microbial inoculation at a specific inoculum density of broth media (in tubes or microtiter plates) containing antibiotics at varying levels—usually doubling dilutions are used (e.g., 1, 2, 4, 8, and 16 µg/mL). The standardized bacterial suspension is typically $1–5 \times 10^5$ CFU/mL. Following incubation at 35 °C, turbidity is recorded either visually or with an automated reader, and the breakpoint concentration established. The lowest concentration of antibiotic that prevented growth represents the MIC. The precision of this method is to be ±1 twofold concentration [3].

Microtiter plates or ready-to-use strips are commercially available with antibiotics ready prepared in the wells. Standard trays contain 96 wells, each containing a volume of 0.1 mL that allows approximately 12 antibiotics to be tested in a range of eight × twofold dilutions in a single tray [4].

A variation on this approach is the agar dilution method where a small volume of suspension is inoculated onto agar containing a particular concentration of antibiotic, when the inoculum has dried the plate is incubated and again examined for zones of growth. With this microdilution testing, the method uses about 0.05–0.1 mL total broth volume and can be conveniently performed in a microtiter format.

14.2.4 Disc diffusion method

Disc diffusion or the Kirby–Bauer test is one of the classic microbiology techniques, and it is still very commonly used. Because of convenience, efficiency, and cost, the disc diffusion method is probably the most widely used method for determining antimicrobial resistance around the world.

A suspension of the isolate (of approximately $1-2 \times 10^8$ CFU/mL) is prepared to a particular McFarland standard, then spread evenly onto an appropriate agar (such as Müller-Hinton agar) in a Petri dish. The agar typically contains (weight/volume) [5]:

- 30.0% beef infusion;
- 1.75% casein hydrolysate;
- 0.15% starch;
- 1.7% agar;
- pH adjusted to neutral at 25 °C.

With the test, the discs are impregnated with various defined concentrations of different antibiotics are placed onto the surface of the agar. A multichannel disc dispenser can speed up placement of the discs. After incubation (16–24 h at 35 °C) zones of growth inhibition around each of the antibiotic discs are measured to the nearest millimeter (as shown in Figure 14.1). A clear circular zone of no growth in the immediate vicinity of a disc indicates susceptibility to that antimicrobial [6]. Using reference tables, the size of zone can be related to the MIC and results recorded as whether the organism is susceptible (S), intermediately susceptible (I), or resistant (R) to that antibiotic [7].

There are a number of critical steps in this approach, such as which medium is used; depth and moisture content of the agar in the plate; incubation conditions; accurate inoculum density; discs must be firmly placed in contact with the agar surface otherwise the diffusion rate will not be correct. The advantages of the disc method are the test simplicity that does not require any special equipment.

Preprepared antibiotic discs with full quality control documentation provided by the manufacturer maintain reproducibility and considerably increases assay reliability. Discs should always be manufactured to an acceptable specification, for example, from the US Food and Drug Administration (FDA), WHO, and DIN. The DIN standard has the tightest range with antibiotic concentrations within 90–125% of that stated.

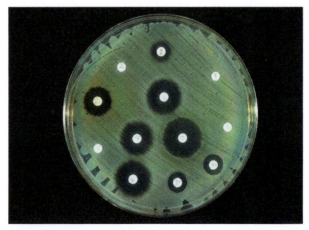

Figure 14.1 Classic disc diffusion assay.
Photograph courtesy of Pharmig.

A variation on this approach is to use a strip impregnated along its length, with a gradient of different concentrations of antimicrobial. This method employs thin plastic test strips that are impregnated on the underside with a dried antibiotic concentration gradient and are marked on the upper surface with a concentration scale. Following incubation this creates an ellipse-shaped zone of no growth, where the ellipse meets the strip, the MIC can be read from the concentration markings on the strip. These are easy to read, no tables need to be referenced to get an MIC value, and the test requires fewer manipulations, as one strip will cover the whole concentration range. These again can be manually or instrument read [8].

14.2.5   Test variability

There are several factors can affect the accuracy of the AST results. These conditions are pH, moisture, effects of medium components, and microbial inoculum.

14.2.5.1   pH

pH of the medium is an important factor that influences the accuracy of the AST. If the pH of the medium is too low than the desired pH, certain drugs, such as aminoglycosides, quinolones, and macrolides, lose their potency. In contrast, antibiotic classes, such as tetracyclines, appear to have excess activity a lower pH, and the vice versa happens in the case of the higher pH.

14.2.5.2   Moisture

The presence of moisture content on the medium can counter act with accuracy of the susceptibility testing. It is important to remove the excess moisture present in the agar surface, by keeping it in the laminar flow hood for few minutes.

14.2.5.3   Effects of medium components

If the media selected for the antibiotic susceptibility contains excessive amounts of thymine or thymidine compounds, they will reversibly inhibit the action of certain antimicrobial agents such as trimethoprim groups. This reversible inhibition yields smaller or less distinct or even no zones and will be misinterpreted as resistant antibiotics. Mueller-Hinton agar is low in thymine and thymidine contents, and it can be used successfully to study the susceptibility of antibiotics. Also, the medium containing excessive cation reduces the zone size, while low cation content results in unacceptably large inhibition zones.

14.2.5.4   Microbial inoculum

The amount of the organism used for the susceptibility testing is standardized using a turbidity standard. This is obtained by a visual approximation using McFarland standard of 0.5, or else it can be determined by using a spectrophotometer with optical density of 1 at 600 nm wavelength.

14.2.6   Automated methods

Disc diffusion and broth dilution techniques can be semiautomated by using image analysers to read zones or turbidity readers for the broths, these give a more objective result and can come with software for automatically interpreting results. Automated systems that are widely used, these normally combine identification with sensitivity testing. The use of instrumentation can standardize the reading of end points and often produce susceptibility test results in a shorter period than manual readings because sensitive optical detection systems allow detection of subtle changes in bacterial growth.

With automated methods, the whole test is set up and read automatically not only is the workload reduced but also the result is less subjective, more reproducible. Results are usually faster, with same day results possible as the instruments monitor growth by taking continuous readings and base results on growth kinetics. While automated systems have many advantages, they can be less flexible in terms of the choice of antibiotics available, consumable costs are usually higher, and equipment costs need to be met whether by outright purchase, leasing or reagent rental deals, together with service and maintenance charges.

For both the semiautomated zone readers and the fully automated ID and susceptibility systems, the data collected can be assessed by expert or smart software systems for interpretation, highlighting unusual anomalous results, suggesting other possible antibiotics to try and can be exported to other laboratory information management systems for further reporting.

The use of genotypic approaches for detection of antimicrobial resistance genes has been promoted as a way to increase the rapidity and accuracy of susceptibility testing. Methods that employ the use of comparative genomics, genetic probes, microarrays, nucleic acid amplification techniques (such as polymerase chain reaction (PCR)), and DNA sequencing are capable of increased sensitivity, specificity, and speed in the detection of specific known resistance genes [9].

14.2.7   Results interpretation

Whichever method is used, the result provides a key cut-off point that equates to the MIC of antibiotic for that test isolate, and methods initially require a pure culture to be prepared, which may take 1–2 days.

With antimicrobial efficacy testing, different products will have different pass criteria. This will be based on the route of administration [10]. However, the general interpretation criteria will be [11]:

- a "susceptible" result indicates that the patient's organism should respond to therapy with that antibiotic using the dosage recommended normally for that type of infection and species;
- an organism with an MIC or zone size interpreted as "resistant" should not be inhibited by the concentrations of the antibiotic achieved with the dosages normally used with that drug;
- an "intermediate" result indicates that a microorganism falls into a range of susceptibility in which the MIC approaches or exceeds the level of antibiotic that can ordinarily be achieved and for which clinical response is likely to be less than with a susceptible strain.

As a note of caution, sometimes the "intermediate" result can also mean that certain variables in the susceptibility test may not have been properly controlled, and that the values have fallen into a "buffer zone" separating susceptible from resistant strains.

While the AST, in its various forms, more often than not produces accurate results, it should be borne in mind that the effectiveness of individual antibiotics varies with the location of the infection, the ability of the antibiotic to reach the site of infection, as well as with the ability of the bacteria to resist or inactivate the antibiotic.

14.3   Antimicrobial efficacy testing (preservative efficacy testing)

The antimicrobial effectiveness test (AET) is used to assess preservative efficacy of products in multi-dose containers. The AET is sometimes referred to also as the preservative effectiveness test. The preservative efficacy of products can either be due to added preservatives, or the inherent properties of the product without the addition of preservatives.

More often, the preservative is added to the product at some point during formulation. If a multiuse pharmaceutical preparation does not itself have adequate antimicrobial activity, antimicrobial preservatives may be added, particularly to aqueous preparations, to prevent proliferation, or to limit microbial contamination which, during normal conditions of storage and use, particularly for multidose containers, could occur in a product and present a hazard to the patient from infection and spoilage of the preparation.

The AET consists of challenging the preparation, wherever possible in its final container, with a prescribed inoculum of suitable microorganisms storing the inoculated preparation at a prescribed temperature, withdrawing samples from the container at specified intervals of time and counting the organisms in the sample removed.

Importantly, the test is not intended to challenge the ability of product in multidose containers to withstand in-use contamination. Instead, the test is designed to show that the product will be stable over a prolonged period of time in relation to microbial contamination. The purpose of the test is, therefore, designed to focus on activity of the preservative systems as a protection against inadvertent contamination during storage and use of the product. The test is used during product development to determine the effectiveness of the product and during stability to demonstrate the preservative system is stable over time. Certain microorganisms can adversely impact (reduce or inactivate) the activity of certain products. It is important to understand the microbial load of the nonsterile finished products in order to determine if the product will react the way it was intended. Likewise, it is important to understand the effectiveness of the preservative system to ensure the product will remain active as intended overtime without garnering microorganisms that can cause human infection [12].

Typical categories of products tested include:

* parenteral and ophthalmic preparations;
* topical preparations;
* oral preparations.

The first contamination-related consideration is with the selection of the preservative and its efficacy. An ideal preservative is a rapidly effective and topically nonirritating. It may be a single antimicrobial agent or a mixture of such agents. Preservatives are designed to prevent the growth or to destroy microorganisms accidentally introduced into the product when the container is opened during use. Preservatives are a necessary additive.

There are several critical considerations in selecting a preservative for inclusion in the dosage form. These include preservative stability, chemical compatibility with the other components of the formulation, compatibility with packaging materials, and concentration. Preservative agents must to be effective throughout the entire shelf life of the product. A preservative will only provide protection from microbial growth for a short time period. For this reason, 28 days is typically stated as the maximum shelf-life after the preservative-containing product is opened [13].

14.3.1 Antimicrobial efficacy test

The compendial AET is essentially a suspension test designed to demonstrate the extent of microbial kill. The AET test comprises a controlled inoculum of the challenge organisms is placed in suspension with the preservative sample to be tested, and the number of survivors determined at different time points. Key aspects of the test are with developing a method for neutralization of the preservative. Residual preservative in the recovery agar could artificially depress the recovery of viable cells. It is, thus, important to neutralize this residual activity to get accurate counts of survivors [14].

The test is not designed to be a quality control release test. To verify effectiveness of the preservative, the AET is assessed during product validation. Obtaining successful validation of the assay can be complex depending on the product. Key factors in developing a proper method include some experimentation as well as knowledge of the test material properties.

The test is not (yet) subject to international harmonization. It is important to note that while the US Pharmacopeia (USP), Japanese Pharmacopoeia (JP), and European Pharmacopoeia (Ph. Eur.) all describe the assay in a similar fashion, but the acceptance criterion varies amongst the different compendial chapters.

The references are:

- USP <51> Antimicrobial Effectiveness Test (AET);
- Ph. Eur. 5.1.3 Efficacy of Antimicrobial Preservation;
- JP 19. Preservatives-Effectiveness Tests (PET);
- ICH Guidance Q6A Specifications: Test Procedures and Acceptance Criteria for New Drug Substances and New Drug Products: Chemical Substances.

With the pharmacopeia chapters, there are some differences with the microorganisms to be used in the tests [15]. These are shown in Table 14.1.

Antimicrobial preservatives are substances that are added to products to limit microbial contamination during normal use or storage. Antimicrobial agents may be harmful to patients. If possible, it is important to demonstrate that the agents are effective in the final package and that they are safe for human use. Products that have

Table 14.1 Comparison of test microorganisms across the main pharmacopeia

Pharmacopeia	Microorganisms
USP <51>	*Candida albicans* ATCC 10231, *Aspergillus brasiliensis* ATCC 16404, *Escherichia coli* ATCC 8739, *Pseudomonas aeruginosa* ATCC 9027, *Staphylococcus aureus* ATCC 6538
JP 19	*Candida albicans* ATCC 10231, *Aspergillus brasiliensis* ATCC 16404, *Escherichia coli* ATCC 8739, *Pseudomonas aeruginosa* ATCC 9027, *Staphylococcus aureus* ATCC 6538
Ph. Eur. 5.1.3	*Candida albicans* ATCC 10231, *Aspergillus brasiliensis* ATCC 16404, *Pseudomonas aeruginosa* ATCC 9027, *Staphylococcus aureus* ATCC 6538 In addition, environmental isolates can be used to substitute any of the above if they are considered to be more appropriate

inherent antimicrobial properties must also be analyzed for antimicrobial effectiveness. Antimicrobial effectiveness must be demonstrated for all products multidose containers.

The AET is used to evaluate the effectiveness of preservative systems in multidose dosage forms. Products in multidose forms that do not contain added preservatives must still comply with the test to demonstrate that the inherent antimicrobial properties of the product are effective.

The test involves inoculating a measured amount of product with known amounts of microorganisms. Whenever possible, the original containers are utilized for the assay. The containers are protected from light and incubated at ambient temperature for 28 days. The death rate is measured over a 28-day period and compared with the acceptance criteria outlined in the compendial guidance documents.

While the method across the three pharmacopeia varies, they are broadly similar. The section below describes the basis of the Ph. Eur. method. Readers who have an interest in other compendial methods are advised to refer to these.

The method involves the following steps:

- prepare suspensions of each microorganism for the study;
- confirm the estimate of the suspensions. Use the plate count or membrane filtration method;
- inoculate a series of containers of the product to be examined, each with a suspension of one of the test microorganisms. The number of containers is left to the user to define; however, this would ordinarily not be fewer than five containers;
- inoculate each container with one inoculum suspension and mix. The volume of inoculum does not exceed 1.0% of the volume of product. Final concentration is between 1×10^5 and 1×10^6 CFU/mL of product;
- incubate the inoculated containers at $22.5 \pm 2.5\,°C$ (protected from light) and sample at the appropriate intervals;
- remove a suitable sample from each container, typically 1 mL or 1 g, at 0 h and at appropriate intervals according to the type of product. Ensure antimicrobial properties are removed by dilution, filtration or an inactivator. Determine the number of viable cells by the plate count or membrane filtration method. The procedure needs to be validated:

- in the membrane filtration method, filtration must be performed with filters that have a pore size not greater than 0.45 μm. The type of filter material is chosen in such a way that the bacteria-retaining efficiency is not affected by the components of the sample. Common filter materials include cellulose, nylon, and polyvinylidene fluoride (PVDF);

- in terms of culture media, tryptic soy agar (TSA) is used for the culturing of bacteria and Sabouraud dextrose agar (SDA) for the culturing of fungi. For the cultivation of the test organisms, the media should contain a suitable inactivator (neutralizer) for the specific antimicrobial properties in the product to the broth and/or agar media used for the test procedure if required. Media used for testing needs to be tested for growth promotion by inoculating the medium with appropriate microorganisms;

- calculate the log reduction.

In terms of the acceptance criteria, the main pharmacopeia again differ. These differences are shown in Table 14.2.

With the table, the pharmacopoeia divide products into different categories (either by description or using a numbering system).

The Ph. Eur. categories are described (as per Table 14.2), with the USP, the categories are:

- Category 1—injections, other parenteral including emulsions, otic products, sterile nasal products, and ophthalmic products made with aqueous bases or vehicles;
- Category 2—topically used products made with aqueous bases or vehicles, non-sterile nasal products, and emulsions, including those applied to mucous membranes;
- Category 3—oral products other than antacids, made with aqueous basesor vehicles;
- Category 4—antacids made with aqueous bases or vehicles.

Products in different categories require testing at different time intervals. Reference to the compendial text should be sought.

14.3.2   Antimicrobial effectiveness test validation

The suitability of the methods to recover microorganisms if they are present must be established through method verification. Method validation is typically performed on three lots of material to demonstrate the robustness of the method.

With the validation exercise:

- inoculate each container with 0.1 mL (or more as determined by the volume of product) of each of the challenge microorganism inoculums prepared and label the containers accordingly. There should be at least one uninoculated product control container;
- the volume of inoculum and product in the sterile containers may be adjusted as long as the volume of the suspension inoculum used is between 0.5% and 1% of the volume of product;
- mix the solution well to ensure a homogenous distribution of the microorganisms. Determine the number of viable microorganisms in each inoculum suspension by referencing the results from the N0 (or titer);
- then, calculate the initial concentration of microorganisms per milliliter of product under test. The target inoculum should yield a suitable concentration between 1×10^5 and 1×10^6 CFU/mL of product.

Table 14.2 Acceptance criteria for the antimicrobial effectives test

Pharmacopeia	Acceptance criteria	Interpretation
USP <51>	Category 1: NLT 1.0 log from initial at 7 days, NLT 3.0 log reduction from the initial count at 14 days, and no increase from the 14 days at 28 days for bacteria. Yeast and moulds—no increase from the initial calculated count at 7, 14, and 28 days Category 2: bacteria—NLT 2.0 log reduction from the initial count at 14 days and no increase from the 14 day counts at 28 days. Yeast and mould—no increase from the initial calculated count at 14 and 28 days Category 3: bacteria—NLT 1.0 log reduction from the initial count at 14 days and no increase from the 14 days count at 28 days. Yeast and moulds—no increase from the initial calculated count at 14 and 28 days Category 4: bacteria, yeast, and moulds—no increase from the initial calculated count at 14 and 28 days	No increase is defined as not more than a 0.5 log unit higher than the previous value measured
JP 19	Category 1A: bacteria: After 14 days 0.1% of inoculum count or less/after 28 days the same level or less after 14 days. Yeasts/moulds: same or less than inoculum count after 14 days/same or less than inoculum count after 28 days Category 1B: bacteria: After 14 days 1% of inoculum count or less/after 28 days the same level or less after 14 days. Yeasts/moulds: same or less than inoculum count after 14 days/same or less than inoculum count after 28 days Category 2: bacteria: same or less than inoculum count at 14 days/same or less than inoculum count at 28 days. Yeast/moulds: same or less than inoculum at 14 days/same or less inoculum count at 28 days Category 1C: bacteria: after 14 days 10% of inoculum count or less/after 28 days the same level or less after 14 days. Yeasts/moulds: same or less than inoculum count after 14 days/same or less than inoculum count after 28 days	When the results are obtained, the product examined is considered to be effectively preserved
Ph. Eur. 5.1.3	Parenteral and ophthalmic preparations: bacteria (A) 6 h = 2 log/24 h = 3 log/28 day = no recover fungi (A) 7 day = 2 log/28 day = no increase/bacteria (B) 24 h = 1 log/7 day = 3 log/28 day = no increase fungi (B) 14 day = 1 log/28 day = no increase Topical preparations: bacteria (A) 2 day = 2 log/7 day = 3 log/28 day = no recover fungi (A) 14 day = 2 log/28 day = no increase/bacteria (B) 14 day = 3 log/28 day = no increase fungi (B) 14 day = 1 log/28 day = no increase Oral preparations: bacteria 14 days = 3 log/28 days = no increase. Fungi: 14 days = 1 log/28 days = no increase	When the results are obtained, the product examined is considered to be effectively preserved. In justified cases where the A criteria cannot be obtained, the B criteria can be used

Note: NLT, not less than.

To assess the log reduction, the following formula should suffice:

$$\begin{aligned}
\log \text{reduction} &= \log \text{of initial calculated CFU/mL} \\
&\quad - \log \text{of product challenge results CFU/mL} \\
&= \log \text{of } 1 \times 10^5 \text{ CFU/mL} \\
&\quad - \log \text{of } 1.0 \times 10^2 \text{ CFU/mL } (\text{or } 100 \text{ CFU/mL}) \\
&= 5 - 2 = 3
\end{aligned}$$

Useful information for the development process includes:

- if the product contains antimicrobial activity, this should be neutralized;
- if inactivators are used, their efficacy and their absence of toxicity for microorganisms must be demonstrated;
- common neutralizing agents and methods include the addition of polysorbate, the addition of lecithin, and/or dilution methods.

During the course of the validation exercise, it is important to note:

- How does the material react to varying scenarios?
- How does the material dissolve?
- How easily/accurately are inoculated microorganisms recovered?
- Does the pH need to be adjusted?
- Which method (membrane filtration or pour plate) is the right method?
- Which neutralizers are needed, if any?

In addition to the above, it should be borne in mind that preservatives are toxic chemicals; therefore, toxicological tests must also be performed.

14.4   Conclusion

This chapter has consider two parts of antimicrobial activity that relate to pharmaceutical products. The first was in relation to antibiotics, used to treat general or specific bacterial infections. Here, although a variety of methods exist, the goal of in vitro antimicrobial = susceptibility testing is the same: to provide a reliable predictor of how a microorganism is likely to respond to antimicrobial therapy in the infected host.

The second area examined in this chapter was with the addition of preservatives to products in order to extent the shelf-life of the product beyond what would be possible from the product ingredients without the addition of the preservative.

Both antibiotics and preservatives are similar in terms of falling under the classification antimicrobials [16]. There are a vast spectra of chemicals that fall within this category demonstrating the effectiveness of these, whether designed to treat, or to preserve, forms an important part of clinical and pharmaceutical microbiology.

References

[1]   Waksman SA. What is an antibiotic or an antibiotic substance? Mycologia 1947;39(5):565–9.
[2]   Sandle T. Sterile ophthalmic preparations and contamination control. J GXP Compliance 2014;18(3):1–5.

[3] Ericsson JM, Sherris JC. Antibiotic sensitivity testing: report of an international collaborative study. Acta Pathol Microbiol Scand 1971;217(Suppl.):1–90.

[4] Jorgensen JH, Turnidge JD. Antibacterial susceptibility tests: dilution and disk diffusion methods. In: Murray PR, Baron EJ, Jorgensen JH, Landry ML, Pfaller MA, editors. Manual of clinical microbiology. 9th ed. Washington, DC: American Society for Microbiology; 2007. p. 1152–72.

[5] Atlas RM. Handbook of microbiological media. London: CRC Press; 2004, p. 1226.

[6] Bauer AW, Kirby WMM, Sherris JC, Turk M. Antibiotic susceptibility testing by a standardized single disk method. Am J Clin Pathol 1966;45:493–6.

[7] Clinical and Laboratory Standards Institute. Performance standards for antimicrobial disk susceptibility tests. Approved standard M2-A10. Wayne, PA: Clinical and Laboratory Standards Institute; 2009.

[8] Jorgensen JH, Ferraro MJ, McElmeel ML, Spargo J, Swenson JM, Tenover FC. Detection of penicillin and extended-spectrum cephalosporin resistance among *Streptococcus pneumoniae* clinical isolates by use of the E test. J Clin Microbiol 1994;32:159–63.

[9] Chen S, Zhao S, McDermott PF, Schroeder CM, White DG, Meng J. A DNA microarray for identification of virulence and antimicrobial resistance genes in *Salmonella* serovars and *Escherichia coli*. Mol Cell Probes 2005;19:195–201.

[10] Jimenez L. Antimicrobial effectiveness test and preservatives in pharmaceutical products. In: Jimmenez L, editor. Microbiological contamination control in the pharmaceutical industry. New York: Marcel-Dekker; 2004. p. 283–99.

[11] Clinical and Laboratory Standards Institute. Development of in vitro susceptibility testing criteria and quality control parameters; approved guideline. CLSI document M23-A. 3rd ed. Wayne, PA: Clinical and Laboratory Standards Institute; 2008.

[12] Sutton S, Porter D. Development of the antimicrobial effectiveness test as USP chapter <51>. PDA J Pharm Sci Technol 2002;56(6):300–11.

[13] United States Pharmacopeia. USP 36-NF 31, antimicrobial effectiveness testing 51. Rockville, MD: USP; 2013, p. 54–5.

[14] Missel PJ, Lang JC, Rodeheaver DP, Jani R, Chowhan MA, Chastain J, et al. Design and evaluation of ophthalmic pharmaceutical products. In: Florence AT, Siepmann J, editors. Modern pharmaceutics—applications and advances. New York: Informa; 2009. p. 101–89.

[15] Moser CL, Meyer BK. Comparison of compendial antimicrobial effectiveness tests: a review. AAPS PharmSciTech 2011;12(1):222–6.

[16] Fassihi RA. Preservation of medicines against microbial contamination. In: Block SA, editor. Disinfection sterilization and preservation. 4th ed. Philadelphia, PA: Lea and Febiger; 1991. p. 871–86.

Cleaning and disinfection

15

15.1 Introduction

Cleaning and disinfection practices are an essential part of contamination control in the pharmaceutical industry, especially in relation to the microbiological control of cleanrooms. Disinfectants are also important for use in the microbiology laboratory.

An important step toward achieving microbial control within a cleanroom is the use of defined cleaning techniques, together with the application of detergents and disinfectants. The objective of cleaning and disinfection is to achieve appropriate microbiological cleanliness levels required and for an appropriate period of time [1].

This chapter examines both detergents (which "clean") and disinfectants (which remove or eliminate microorganisms). Detergents are cleaning agents and are deployed to remove "soil" from a surface. The removal of soil is an important step prior to the application of a disinfectant, for greater the degree of soiling remaining on a surface, then the lesser the effectiveness of disinfection. A disinfectant is a type of chemical germicide that is capable of eliminating a population of vegetative microorganisms (although some disinfectants are sporicidal, a chemical does not need to be sporicidal to be classified as a disinfectant). A disinfectant that can kill spores is sometimes described as a sterilant or chemosterilant [2]. Disinfectants, of varying formulations, have been used since the late nineteenth century [3].

Disinfectants vary in their effectiveness against different types of microorganisms, a variation relating to both the intrinsic resistance of different microorganisms and the range of different types and formulations of disinfectants. Furthermore, different disinfectants act in different ways depending upon their active ingredients.

15.2 Cleaning

Cleaning is the process of removing residues and "soil" (such as dirt, grease, and protein residues) from surfaces to the extent that they are visually clean. This involves defined methods of application and often the use of a detergent. Importantly, the act of cleaning is necessary prior to the application of a disinfectant for a surface needs to be properly cleaned before the application of a disinfectant in order for the disinfectant to work efficiently [4], as disinfectants can either be inactivated by organic residues or the soil can create a barrier which prevents the disinfectant from reaching all of the microbial cells.

While "cleaning" is not "disinfection", the cleaning process can remove or dilute microbial populations. Furthermore, many detergents have chemical additives that can "disinfect." However, a cleaning agent will not meet the criteria for disinfection required by the European and US standards for disinfectant validation in terms of reducing a microbial population of a defined range by the required log reduction.

Pharmaceutical Microbiology. http://dx.doi.org/10.1016/B978-0-08-100022-9.00015-3

The act of cleaning normally requires the use of a detergent. A detergent is a chemical used to clean equipment or surfaces by removing unwanted matter (soil). Detergents generally work by penetrating soil and reducing the surface tension (which adhere soil to the surface) to allow its removal (in crude terms, a detergent increases the "wettability" of water). Many detergents are synthetic surfactants (an acronym for surface active agents).

Surfactants are schizophrenic molecules that have two sides to their nature. One part is solvent-loving or lyophilic (hydrophilic), and another is solvent-hating or lyophobic (hydrophobic). Surfactants remove particles from surfaces by either capillary effects or electrostatic forces (many detergents contain differently charged ions that can cause microorganisms to repel each other). This repulsion causes the microorganisms to disassociate from the surface and become suspended. Suspended microorganisms (planktonic state) are easier to remove from the surface by the rinsing effect of the detergent (or a subsequent water rinse) or to be destroyed by the application of a disinfectant [5].

There are two key considerations when selecting detergents.

1. The chemical composition of the detergent.

With the chemical nature of the detergent, it is typical that detergents are neutral and non-ionic solutions. Furthermore, it is preferred that the detergents used are low- or nonfoaming.

2. The compatibility of the detergent with the disinfectant.

In terms of compatibility, it is important that any detergent used should be compatible with the disinfectants used, for some detergents can leave residues that can neutralize the active ingredient in certain disinfectants thereby reducing the microcidal properties of the disinfectant [6].

15.3 Disinfection

A disinfectant is a chemical agent, one of a very diverse group of products, which reduces the number of microorganisms present either by removing or destroying them. In literature, various terms are applied in relation to this activity: disinfectant, antiseptic, asepsis and sanitizer.

The term disinfection is normally applied to an inanimate object (sometimes the term biocide is used, although this relates to a larger group of chemical agents). The term antiseptic is used to describe the reduction of a microbial population on living tissue [7]. Thus, an antiseptic is a disinfectant that can safely be applied to the surface of the skin (sometimes the terms "hand sanitizer" or "hand disinfectant" are used interchangeably) [8]. In turn, the term disinfectant is usually reserved for liquid chemical germicides, which cannot be applied to tissues because of their corrosive or toxic nature [9].

Asepsis can relate to the use of disinfectants to disinfect an area such as an operating theater (and the term is more commonly associated with healthcare) [10]. It is a separate term to "aseptic technique," which, in the laboratory sense, relates to avoiding personnel contamination of devices intended to be sterile [11].

The term sanitizer is open to different interpretations. Within Europe, it is normally taken to be an agent that both cleans and disinfects (normally a disinfectant that contains a cleaning agent). Within North America (defined by the US Environmental Protection Agency), however, the term is normally applied to an antimicrobial agent for use on nonfood contact surfaces. Sanitization is a general description for reducing a microbial population. Disinfection is a more precise term, as it can be related to the requirements of international standards in relation to the requirement that the chemical agent must reduce a known number of microorganisms (a property demonstrated through validation).

15.3.1 Disinfectant efficacy

There are a number of important criteria that affect the performance and efficacy of disinfectants. These factors are:

(i) Concentration

Disinfectants are manufactured or validated to be most efficacious at a set concentration range (the proportion of the chemical to water). The setting of this concentration range involves ascertaining the minimum inhibitory concentration (MIC). The MIC is the lowest concentration of the disinfectant that is shown to be bacteriostatic or bactericidal under experimental conditions. Experimental conditions are normally based on the examination of a disinfectant solution in suspension in the absence of soil. The MIC is measured through kinetic studies of the dilution coefficient. Kinetic studies demonstrate the effect of a change in concentration against cell death rate over time. The higher a disinfectant's concentration exponent, the longer it takes to kill cells. For example, if a disinfectant with a set concentration exponent was diluted by a factor of 2, the time taken for it to kill cells comparatively would double [12]. The MIC is normally set by the manufacturer of the disinfectant.

(ii) Time

Time is an important factor in the application of disinfectants for two reasons: in relation to the contact time of the disinfectant and the expiry time of the disinfectant solution. Contact time (sometimes called the dwell time) is the time taken for the disinfectant to bind to the microorganism, traverse the cell wall and to reach the specific target site for the disinfectant's particular mode of action. Many disinfectants work best and meet product label claims when allowed to work for several minutes before wiping or rinsing.

Contact time relates to the concentration of the disinfectant (variation to the concentration of a disinfectant may alter the contact time required). In practical situations, there are many variables which can alter the contact time. These include the type, concentration, and volume of the disinfectant; the nature of the microorganisms; the amount and type of material present that is likely to interfere with the active ingredient; the temperature of the disinfectant; and the surface that the disinfectant is applied to.

Another aspect relating to time is the deterioration of a disinfectant solution over time. This is more important where a solution of disinfectant is prepared "in-house" from a concentrate than to ready-prepared solutions, which have been validated by the manufacturer and will come with an assigned expiration time. For ready-prepared solutions, an expiry time limit for the disinfectant solution should be established through chemical testing. As a rule, fresh solutions of a disinfectant should be used for each application and between cleanrooms.

(a) Number

An antimicrobial agent, like a disinfectant, is considerably more effective against a low number of microorganisms than a higher number or a population with a greater cell density. Similarly, a disinfectant is more effective against a pure population than mixed grouping of microorganisms. A routine disinfectant procedure will be unlikely to kill all microorganisms present, and a number will remain viable. Whether the surviving microorganisms multiply in sufficient number is dependent upon the conditions in which the surviving population remains, the available nutrients and the time between repeat applications of the disinfectant.

(b) Type of microorganism and resistance

Different types of microorganism have varying levels of resistance to broad spectrum disinfectants as Figure 15.1 shows. The increased resistance shown is primarily due to the cell membrane composition or type of protein coat.

The hierarchy of microorganisms in Figure 15.1 is placed in order of resistance. Resistance is either due to the natural genetic properties of the microorganisms (intrinsic), as shown in Figure 15.1, or it is acquired through phenotypic (organism's actual observed properties, such as colony pigment) or genotypic (genetic) variations (similar to antibiotic resistance, through the over-use of one type of disinfectant. However, This form of "acquired resistance" is contentious). Generally, innate sensitivity results in

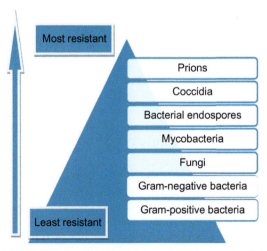

Figure 15.1 Sensitivity of different microorganisms to broad spectrum disinfectants.
© Tim Sandle.

Gram-negative bacteria being more resistant to disinfectant applications than Gram-positive bacteria, based on the composition of their cell wall. In turn, for bacteria, endospores are the most resistant because of the relative impermeability of the spore coat.

(c) Location of microorganisms

The location of microorganisms influences the effectiveness of disinfectant treatment. Microorganisms in suspension are easier to kill than those affixed to surfaces. This is due to the mechanisms of microorganism attachment, such as bacteria fixing themselves using fimbriae or when a biofilm community develops. A biofilm is a complex aggregation of microorganisms growing on a solid surface. Bacteria living in a biofilm can have significantly different properties from free-floating bacteria, as the dense and protected environment of the film means that they have increased resistance to detergents and antibiotics, as the dense extracellular matrix and the outer layer of cells protect the interior of the community.

Thus, such positioning can impact upon the contact time required for the disinfectant to bind to the microorganism, cross the cell wall and act at the required site. This is why, with disinfectant validation, the surface test is regarded as more meaningful and robust than the suspension test.

(iii) Temperature and pH

Each disinfectant has an optimal pH and temperature at which it is most effective. If the temperature or pH is outside this optimal range then the rate of reaction (log kill over time) is affected.

Temperature influences the rate of reaction. Most disinfectants are more effective and kill a population faster at higher temperatures although many disinfectants, due to practical considerations relating to cleanroom use, are manufactured to be used at ambient conditions. Disinfectants that are sensitive to temperatures other than at ambient are normally assessed using a temperature coefficient, or Q_{10} (which relates the increase in activity to a 10 °C rise in temperature) [13].

The effect of pH is important because it influences the ionic binding of a disinfectant to a bacterial cell wall thereby ensuring disinfectant molecules bound to a high number of microorganisms. Many disinfectants are more stable at a set pH range. The use of a disinfectant outside of its desired pH range results in reduced efficacy.

(iv) Interfering substances

The presence of different substances on the surface or in the equipment requiring disinfection can affect the efficacy of the disinfectant in various ways ranging from increasing the contact time to complete inactivation. In order for a disinfectant to be effective, it must come into contact with the microbial cell and be absorbed into it. If substances (or organic load), such as oil, dirt, blood serum, protein, food body waste paper or grease, act as a barrier between the microbial cell and the disinfectant, the efficacy of the disinfectant will be adversely affected. The presence of such substances (soil) halts disinfectant efficacy by either reacting with the disinfectant or creating a barrier for the disinfectant. This effect is increased if the surface itself has defects and crevices, which limit disinfectant penetration [14]. When purchasing disinfectants, it is important to note if the label claim indicates if the disinfectant remains effective in the presence of small amounts of organic matter.

(v) Water

Many disinfectants do not work well in hard water. For use in the pharmaceutical and healthcare sectors, disinfectants are normally prepared using deionized or demineralized water (or water for injections in the higher grade cleanrooms).

15.3.2   Types of disinfectants

There are various types of disinfectants available. Different types of disinfectants have different spectra of activity, modes of action and differing efficacies. Some disinfectants are classed as bacteriostatic, where the ability of the bacterial population to grow is halted. Here the disinfectant can cause selective and reversible changes to cells by interacting with nucleic acids, inhibiting enzymes, or permeating into the cell wall. Once the disinfectant is removed from contact with bacteria cells, the surviving bacterial population could potentially grow.

Other disinfectants are bactericidal in that they destroy bacterial cells through different mechanisms. These include causing structural damage to the cell, autolysis, cell lysis, or by the leakage or coagulation of the cytoplasm. Within these groups, the spectrum of activity varies, with some disinfectants being effective against vegetative Gram-positive and Gram-negative microorganisms only, while other disinfectants are effective against fungi. Other disinfectants have a broader spectrum and are sporicidal in that they can cause the destruction of endospore-forming bacteria. However, a chemical agent does not have to be sporicidal in order to be classed as a "disinfectant" [15]. The bacteriostatic, bactericidal, and sporicidal properties of a disinfectant are influenced by many variables.

There are a number of disinfectants available in the market with different modes of activity and of varying effectiveness against microorganisms. There are various approaches to the categorization and sub-division of disinfectants including grouping by chemical nature, mode of activity or by microstatic and microcidal effects on microorganisms. The two principal categories are the division into oxidizing and nonoxidizing chemicals.

The majority of nonoxidizing disinfectants have specific modes of action against microorganisms, but generally they have a lower spectrum of activity compared to oxidizing disinfectants [16]. The most common types of nonoxidizing disinfectants are alcohols, quaternary ammonium compounds, and phenolics. With oxidizing disinfectants, these chemicals generally have nonspecific modes of action against microorganisms. They have a wider spectrum of activity than nonoxidizing disinfectants, with most types able to damage endospores, but they pose greater risks to human health. Examples include halogens, peracetic acid, and hydrogen peroxide.

15.3.3   Selecting disinfectants

Deciding the types of disinfectants to be used represents an important decision. Key criteria include [17]:

(a) A disinfectant must have a wide spectrum of activity

The spectrum of activity refers to the properties of a disinfectant being effective against a wide range of vegetative microorganisms including Gram-negative and Gram-positive bacteria. For in cleanrooms, disinfectants should be bactericidal (that is rather than simply inhibiting microbial growth, it should be capable of killing bacteria). A separate decision to be made is whether the disinfectant is required to be sporicidal. Furthermore, in some facilities the disinfectants should also be virucidal.

(b) A disinfectant should have a fairly rapid action

The speed of action depends upon the contact time required for the disinfectant to destroy a microbial population. Thus, the "minimum" contact time is the time required for the disinfectant to be effective after its application. The contact time is sometimes referred to as the "action time" or "dwell time." Given the requirements of most pharmaceutical manufacturers and healthcare facilities, a disinfectant should ideally have a contact time of 10 min or less, although certain sporicidal disinfectants can require longer contact times.

(c) Disinfectants should not be neutralized by residual matter

Although detergents and effective cleaning practices can remove the majority of soil, including organic matter, some traces will remain. It is important that these organic residues do not interfere with the active ingredient of the disinfectant and reduce its efficacy.

(d) Environmental conditions

Some disinfectants require certain temperature and pH ranges in order to function properly. One type of disinfectant, for example, may not be effective in a cold room due to the lower temperature. The reason for this is the validation standards for disinfectants measure the bactericidal activity at 20 °C; therefore, the disinfectant may not be as effective at higher or lower temperatures.

(e) Non-corrosive

Disinfectants should be noncorrosive. If the disinfectant causes extensive abrasion of a surface, it will either degrade the material or cause cracks and recesses that can harbor microorganisms. It is recognized, however, that the most efficacious disinfectants, especially those that are sporicidal, through repeated applications over time will cause some corrosion. A postdisinfection step to remove disinfectant residues, such as a sterile water rinse or wiping with a milder disinfectant such as 70% isopropyl alcohol (IPA), can minimize material surface damage.

(f) Operator safety

Many disinfectants are toxic or irritant and unpleasant for staff to use. Consideration must be given to safety requirements, material safety data sheets, label information, the toxicity upon human health, and to the protective measures required for staff to use them (such as avoiding contact with exposed skin or the need to use a disinfectant in a well-ventilated area).

(g) Compatibility with the surface to be disinfected

Certain disinfectants may be less effective with certain materials or may cause excessive damage to certain materials, such as the reaction of chlorine dioxide against stainless steel.

(h) Compatibility with detergents used

As discussed above, it is important that the disinfectant and detergent are compatible and that detergent residues do not inactivate the active ingredients in the disinfectant solution.

(i) Residual activity of the disinfectant

Residual activity of the disinfectant may lead to resistant strains or cause problems when an alternative disinfectant is applied. It is a good practice to remove disinfectant residues with a water rinse.

(j) Sporicidal properties

If isolates from the environmental-monitoring program include the recovery of endospore-forming bacteria on a frequent basis or in high numbers, then the use of a sporicidal disinfectant is essential, with the frequency determined by a review of the environmental monitoring program.

(k) Range of formats available

The cleanroom facility will require a disinfectant to be available in several formats. For example, a type and formulation of disinfectant may be required in a ready-to-use format, as a concentrate, or an impregnated wipe, and so on.

(l) Cost

The calculation of cost needs to include not just the price of the disinfectant but also other cost factors such as the time taken to prepare or apply the disinfectant, protective clothing requirements, wastage, and the steps needed for the removal of residues.

(m) Health and safety

The safety aspects of a disinfectant are an important consideration and standard operating procedures should contain appropriate health and safety requirements for using detergents and disinfectants. This should include reference to appropriate personal protective equipment. In particular, contact to eyes, skin and mouth is to be avoided. Safety data sheets must be examined for all disinfectants and detergents and appropriate measures taken to ensure that they are applied properly, in well-ventilated areas.

15.4   Good manufacturing practice requirements

An effective cleaning and disinfection program in pharmaceutical grade areas of a good manufacturing practice (GMP) facility is critical to assure the quality of the products. The use of detergents and disinfectants, and the need to keep cleanrooms clean, is a regulatory requirement within the pharmaceutical sector. The main regulatory documents relating to the use of disinfectants in pharmaceutical manufacturing are:

- FDA Code of Federal Regulations: 21 CFR 211.56b and 21 CFR 211.56c (which refer to sanitation); CFR 211.67 (which refer to equipment and maintenance); CFR 211.182 (which describes the need for a cleaning program); and CFR 211.113b;
- FDA Aseptic Processing Guide, revised 2004;
- USP (General Chapter <1072> Disinfectants and Antiseptics);
- Annex 1 to the EU Guide to Good Manufacturing Practice.

For example, the 21 CFR 211.67 states:

That surfaces and equipment must be "....cleaned, maintained, and sanitized at appropriate intervals to prevent malfunctions or contamination that would alter the safety, identity, strength, quality, or purity of the drug product beyond the official or other established requirements."

To meet regulatory expectations in the pharmaceutical industry, a cleaning and disinfection program is required. The program should consist of a policy, outlining the objectives and the criteria for the selection of materials and cleaning agents; and a procedure, detailing how cleaning is undertaken, along with the techniques and cleaning frequencies. There is an expectation that such programs are regularly reviewed and reflect any changes to cleanroom design and respond to changes to environmental monitoring data.

Although there are some differences between the regulations, a number of similar areas are covered. In summary, the regulations require [18]:

- the need to have written procedures (CFR/EU GMP);
- responsibilities for cleaning should be assigned (CFR). Often this is interpreted as the need to have independent cleaning staff separate from those involved in product manufacture;
- staff must be trained in cleaning techniques and have a training record (CFR/EU GMP);
- details of cleaning frequencies, methods, equipment, and materials must be recorded in written procedures (CFR). This may relate to an approved supplier specification;
- the cleaning of equipment and materials must take place at regular intervals (CFR);
- in designing a disinfectant protocol for the sanitization of floors, walls and surfaces, a pharmaceutical organization will normally select two or three disinfectants for the same application. This is a requirement of regulatory bodies and the strongest pressure for it has come from Europe with the EU GMP guideline stating that "where disinfectants are used, more than one type should be employed" (Annex 1, paragraph 3714). This is normally interpreted as the need for disinfectant rotation (which is discussed below).
- Disinfectants should be rotated (EU GMP/FDA warning letters).
- Inspection of equipment for cleanliness before use should be part of routine operations (CFR).
- A cleaning log should be kept. The purpose is to keep a record of the areas cleaned, agents used and the identity of the operator (CFR).
- The microorganisms isolated (the "microflora") from environmental monitoring programs should be examined for resistant strains (EU GMP). The inference here is that such isolates are incorporated into disinfectant efficacy studies (see Chapter 10).
- The monitoring for microbial contamination in disinfectant and detergent solutions should be periodically undertaken (EU GMP).
- The storing of disinfectant and detergent solutions should be for defined (and short) periods (EU GMP).
- Disinfectants and detergents used in Grades A and B cleanrooms should be sterile before use (EU GMP).
- Room use should be recorded after each operation (CFR/EU GMP).
- Disinfectants should be "qualified" (validated) (CFR).
- There should be a technical agreement with the company who supplies the disinfectant. Ideally the disinfectants purchased should be lot tracked (EU GMP).

The most important GMP consideration in relation to Europe is, arguably, rotation. To conform with GMP expectations, pharmaceutical manufacturer is expected to use at least two disinfectants with different modes of activity [19]. With other territories, the requirement for rotation is not so exact.

While the phenomenon of microbial resistance is an issue of major concern for antibiotics, there are few studies that support development of resistance to disinfectants. The unlikelihood of resistance developing is particular when disinfectants are applied to dry environments, such as cleanrooms, for microbial replication, as the primary process for gaining resistance, is minimal. However, rotation is generally necessary in order to pass a European GMP inspection.

Policies for determining the frequency of rotation vary widely. Some facilities have adopted an even rotation (such as alternating between disinfectants daily or weekly) while others rotate at different frequencies, sometimes to an extreme of 3 months for one disinfectant against 1 week for the alternate disinfectant. Other companies build up a case for only using one disinfectant on a day-to-day basis with a second used very infrequently. In this latter example, it is argued that two disinfectants only need to be employed if the environmental-monitoring data indicate excursions from set limits, and therefore, the inference is that the primary disinfectant is not controlling surfaces [20].

The frequency of rotation needs to be defined by the user, and supporting data can be supplied through field trials. When established it is necessary to continue the detergent application between the changes of disinfectant types in order to remove residues. Once set, there may be a requirement to vary the frequencies of use, such as in response to an increase to microbial counts and as part of a formal investigation into a microbiological data deviation.

15.5   Measuring disinfection effectiveness: Environmental monitoring

To ensure the effectiveness of a cleaning and disinfection program, microbiological environmental monitoring of surfaces and equipment is necessary. The primary methods for conducting these tests involve the use of cotton swabs (with a recovery diluent and later plating onto agar or dissolving prior to membrane filtration) and contact plates or other surface agar techniques. The recovery efficiency of contact plates is generally superior to that of swab. The agars used should contain appropriate neutralizing agents in order to eliminate any disinfectant residues and, thus, allow any recovered microorganisms to grow (this is discussed further in Chapter 10). An appropriate general agar, such as soya-bean casein digest medium, is normally recommended. This agar, onto which swabs are sub-cultured or used in the contact plates, should have a dual incubation step designed to pick up a range of environmental microorganisms. A typical regimen would be 20–25 °C for 48 h followed by 30–35 °C for 72 h.

One way of assessing the effectiveness of a disinfectant is through a field trial [20]. The purpose of field trials is to test a disinfectant's efficacy in practical in-use conditions: the working environment. Most researchers consider that a field trial is the only accurate test of a disinfectant, given the problems with repeatability and reproducibility associated with laboratory-based tests. It is also because the trials examine a selected concentration on surfaces and equipment after the cleaning step has been applied. Field trials have an advantage because they test the disinfectant against a wide range of surfaces and with all the different types of interfering substances that may be present, as well as different physico-chemical conditions, such as temperature and pH.

15.6   Disinfectant efficacy

Disinfectants need to be assessed through disinfectant efficacy studies. Unfortunately, there is no clear-cut approach for this, with differing international standards. The following organizations publish disinfectant validation standards.

- ASTM (American Society for Testing Materials) (ASTM E2614-08 Standard Guide for Evaluation of Cleanroom Disinfectants).
- AOAC (Association of Official Analytical Chemists International) (referenced below).
- CEN (European Standards) (referenced below).
- TGA (Australian Therapeutic Goods Administration).
- USP (US Pharmacopeia) (chapter <1072>).

The most widely used methods are the European CEN standards and the US AOAC standards. The European approach consists of suspension tests, surface tests, and field trials; whereas the emphasis in North America is strongly on surface testing (the hard surface carrier method). Here the carrier test differs slightly in methodology to the European surface test. It is generally regarded that the surface tests are more rigorous than the suspension tests [21].

With suspension tests, a test suspension of bacteria or fungi is added to a prepared sample of the disinfectant under test in simulated "clean" and "dirty" conditions. After a specified contact time, an aliquot is taken, and the bactericidal/fungicidal action is immediately neutralized by the addition of a proven neutralizer (as identified in the basic suspension test). Following this, the number of surviving microorganisms in each sample is determined, and the reduction in viable counts is calculated (expressed in logarithms to base 10).

With surface tests (or "carrier tests"), representative manufacturing surface samples are inoculated with a selection of microbial challenge organisms. A disinfectant is applied to the inoculated surfaces and exposed for a predetermined contact time after which the surviving organisms are recovered using a qualified disinfectant-neutralizing broth and test method (surface rinse, contact plate, or swab). The number of challenge organisms recovered from the test samples (exposed to a disinfectant) is compared with the number of challenge organisms recovered from the corresponding control sample (not exposed to a disinfectant) to determine the ability of the disinfectant to reduce the microbial bioburden. Successful completion of the validation qualifies the disinfectant evaluated for use [22].

Prior to initiating disinfectant efficacy validation, a comprehensive survey of the materials comprising the room surfaces (floors, walls, windows) and equipment (stainless steel, acrylic, vinyl) present in the facility that could potentially be exposed to the disinfectant should be conducted. The use of different surfaces is important because the rates of inactivation on microorganisms on different surfaces can vary considerably. One study demonstrated that bactericidal activity reduced on polyvinyl chloride (PVC) compared with stainless steel. This was a factor both of the material type and the surface conditions, such as the number of pores or ridges. Surfaces of the material can also differ depending upon the degree of finishing with smoother surfaces, like stainless steel or Formica, giving greater repeatability and reproducibility.

15.7   Conclusion

Microbial control in cleanrooms and laboratories is of great importance, and to achieve this use of defined cleaning techniques, together with the application of detergents and disinfectants, is of great importance.

This chapter has examined the basics of cleaning and disinfection and has outlined the key requirements for the selection and use of such agents. In doing so, the chapter has introduced some of the key GMP concepts. The chapter has also outlined the important aspects for the qualification of disinfectants through disinfectant efficacy studies. Of these, the surface studies are the most important. That said, the ultimate assessment of the suitability of a disinfectant is established through a field trial where environmental monitoring data can be evaluated in order to set disinfectant frequencies and the order of disinfectant rotation.

References

[1]   Sutton SVW. Disinfectant rotation in a cleaning/disinfection program for cleanrooms and controlled environments. In: Manivannan G, editor. Disinfection and decontamination: principles, applications and related issues. Boca Raton, FL, USA: CRC Press; 2008. p. 165–74.

[2]   Gorman S, Scott E. Chemical disinfectants, antiseptics and preservatives. In: Denyer SP, Hodges NA, Gorman SP, editors. Hugo and Russell's pharmaceutical microbiology. 7th ed. Oxford: Blackwell; 2004. p. 285–305.

[3]   Worboys M. Public and environmental health. In: Bowler PJ, Pickstone JV, editors. The Cambridge history of science. The modern biological and earth sciences, vol. 6. Cambridge: Cambridge University Press; 2009.

[4]   Bessems E. The effect of practical conditions on the efficacy of disinfectants. Int Biodeter Biodegr 1998;41:177–83.

[5]   Farn RJ. Chemistry and technology of surfactants. London: John Wiley & Sons; 2006.

[6]   Collier L, Balows A, Sussman M. Topley and Wilson's microbiology and microbial infections volume 2: systematic bacteriology. 9th ed. New York: Arnold; 1998, p. 149-70.

[7]   Marcovitch H. Black's medical dictionary. 42nd ed. London: A&C Black; 2010, p. 193.

[8]   Taylor DJ, Green NPO, Stout GW. Biological science 2: systems, maintenance and change. 3rd ed. Cambridge: Cambridge University Press; 1997.

[9]   Brooker C, editor. Mosby's dictionary of medicine, nursing and healthcare professions. Edinburgh: Elsevier; 2010.

[10] Gariepy TJ. Antisepsis. In: Heilbron JL, editor. The Oxford companion to the history of modern science. Oxford: Oxford University Press; 2003. p. 31.

[11]   Wilson J. Infection control in clinical practice. 2nd ed. London: Bailliere Tindall; 2006.
[12]   Russell AD. Assessment of sporicidal efficacy. Int Biodeter Biodegr 1998;41:281–7.
[13]   Cooper MS. Biocides and preservatives. Microbiol Update 2000;18(3):1–4.
[14]   Frank JF, Chmielewski R. Influence of surface finish on the cleanability of stainless steel. J Food Prot 2001;64(8):1178–82.
[15]   O'Leary WM. Practical handbook of microbiology. 2nd ed. Boca Raton, FL: CRC Press; 1977.
[16]   Denyer SP, Stewart GSAB. Mechanisms of action of disinfectants. Int Biodeter Biodegr 1998;41:261–8.
[17]   Sandle T. Cleaning and disinfection. In: Sandle T, editor. The CDC handbook: a guide to cleaning and disinfecting cleanrooms. Surrey, UK: Grosvenor House Publishing; 2012. p. 1–31.
[18]   Sartain EK. Regulatory update: rotating disinfectants in cleanrooms: avoid going in circles. A2C2 Mag 2005;8(3):32–3.
[19]   Sandle T. Practical selection of cleanroom disinfectants. Hosp Pharm Eur 2012;(63):39–41.
[20]   Sandle T. Validation of disinfectants. In: Sandle T, editor. The CDC handbook: a guide to cleaning and disinfecting cleanrooms. Surrey, UK: Grosvenor House Publishing; 2012. p. 241–61.
[21]   Vina P, Rubio S, Sandle T. Selection and validation of disinfectants. In: Saghee MR, Sandle T, Tidswell EC, editors. Microbiology and sterility assurance in pharmaceuticals and medical devices. New Delhi: Business Horizons; 2011. p. 219–36.
[22]   Sandle T. Application of disinfectants and detergents in the pharmaceutical sector. In: Sandle T, editor. The CDC handbook: a guide to cleaning and disinfecting cleanrooms. Surrey, UK: Grosvenor House Publishing; 2012. p. 168–97.

Cleanrooms and environmental monitoring

16

16.1 Introduction

This chapter examines the key aspects relating to the pharmaceutical-manufacturing environment and to matters relating to personnel working in such environments. The chapter examines the design, control, and monitoring of cleanrooms and clean zones. Pharmaceutical manufacturing, both sterile and nonsterile processing, requires that products are developed and manufactured in areas which minimize the potential for contamination through the control of environmental cleanliness and in minimizing the possibility of personnel introducing contamination into the process.

The pharmaceutical-manufacturing environment is based around a series of rooms with specially controlled environments. These are termed cleanrooms. A cleanroom or zone, on one level, is simply a room that is clean. The key aspect, however, is that the level of cleanliness is *controlled*. A cleanroom is a room designed and operated to control particulate levels.

A more specialized meaning is as defined in the international cleanroom standard, ISO 14644-1:

> *A room with control of particulates and set environmental parameters. Construction and use of the room is in a manner to minimise the generation and retention of particles. The classification is set by the cleanliness of the air.*

The regulatory requirements for cleanrooms are detailed by EU GMP or the FDA guidelines. The way in which cleanrooms are qualified and assessed is by a series of ISO standards, such as the ISO 14644 group (which refer to physical parameters, including particle counts); of these, Part 1 (ISO 14644-1) sets the general standard for the classification of air cleanliness and Part 2 (ISO 14644-2) sets out the specifications for testing, are the most important. In addition, ISO 14698 describes some of the standards and testing requirements for biocontamination control.

Cleanrooms can be designed to minimize particulate risk and most cleanrooms operate well, until personnel enter them. People in cleanrooms are the biggest source of contamination [1]. It, therefore, follows that any pharmaceutical process that involves people presents a contamination risk [2].

This chapter discusses the classification and certification of cleanrooms and then proceeds to consider the monitoring steps necessary to demonstrate on-going compliance.

Pharmaceutical Microbiology. http://dx.doi.org/10.1016/B978-0-08-100022-9.00016-5

16.2   Cleanroom contamination

By prescribing a grade or a class to a cleanroom, the areas are then regarded as "controlled" environments. A controlled environment is any area in an aseptic process system for which airborne particulate and microorganism levels are controlled to specific levels to the activities conducted within that environment.

To give this a different perspective, the ambient air outside in a typical urban environment might contain as many as 35,000,000 particles per cubic meter, $0.5\,\mu m$ and larger in diameter, corresponding to an ISO 14644-1 cleanroom class of 9.

The measurement of airborne particle counts is a key part of environmental control. Particles are measured using optical particle counters, and the regulatory requirement is for two sizes of particle to be counted. These are $0.5\,\mu m$ (which is close to the size of a microorganism) and $5.0\,\mu m$ (which is close to the size of a skin cell, which may carry bacteria). These particles are very small and are not visible to the human eye. Generally, the "complete particle" (microorganism in association with the "carrier") is $12\,\mu m$ diameter or larger. Most microorganisms in cleanroom air are attached to dust or to skin flakes or to water droplets [3].

Thus, the particles measured may be nonviable or viable, but because of the association with microorganisms and the assumption that some particles will be microorganisms designing facilities to minimize the number of particles and then monitoring of particulate levels is an important part of contamination control [4].

These levels of cleanliness are established through the design and construction of the cleanroom, particularly [5]:

- The air entering a cleanroom from outside is filtered to exclude dust, and the air inside is constantly re-circulated through HEPA filters. This is controlled through a HVAC (Heating, Ventilation and Air Conditioning) system. The most important part of this is with airfiltration through a HEPA (High Efficiency Particulate Air) filter, or higher grade ULPA (Ultra Low Penetration Air) filters.
- Staff enter and leave through airlocks (sometimes including an air shower stage), and wear protective clothing such as hats, face masks, gloves, boots, and cover-alls.
- Equipment inside the cleanroom is designed to generate minimal air contamination. There are even specialized mops and buckets. Cleanroom furniture is also designed to produce a low amount of particles and to be easy to clean.
- Common materials such as paper, pencils, and fabrics made from natural fibers are often excluded, however, alternatives are available. Low-level cleanrooms are often not sterile (i.e., free of uncontrolled microbes), and more attention is given to airborne particles. As indicated above, particle levels are usually tested using a Laser particle counter.
- Some cleanrooms are kept at a higher air pressure to adjacent (less clean) areas so that if there are any leaks, air leaks out of the chamber instead of unfiltered air coming in.
- Cleanroom HVAC systems also control the humidity to low levels, such that extra precautions are necessary to prevent electrostatic discharges.

Therefore, cleanrooms are designed to minimize and to control contamination. There are many sources of contamination. The atmosphere contains dusts, microorganisms, condensates, and gases. People, in clean environments, are the greatest contributors to contamination emitting body vapors, dead skin, microorganisms, skin oils, and so on. Manufacturing processes will produce a range of contaminants. Wherever there is a process which grinds, corrodes, fumes, heats, sprays, turns, etc., particles and fumes are emitted and will contaminate their surroundings. Another key contamination source is water. This is a continuing problem as water is the main ingredient in pharmaceuticals.

16.3   Cleanroom classification

Cleanrooms and zones are typically classified according to their use (the main activity within each room or zone) and confirmed by the cleanliness of the air by the measurement of particles [6]. Cleanrooms are used in several industries including the manufacture of pharmaceuticals and in the electronics industry. For pharmaceutical cleanrooms, air cleanliness is either based on EU GMP guidance for aseptically filled products, and the EU GMP alphabetic notations are adopted (Table 16.1); or by using the International Standard ISO14644 (Table 16.2), where numerical classes are adopted.

Thus, Grade A is the highest grade (that is the "cleanest"), and Grade D is the lowest (that is the least "clean"). With ISO, the lower the number (such as "5") the

Table 16.1   **EU GMP the typical room uses and associated grades**

Grade	Room use
A	Aseptic preparation and filling (critical zones under unidirectional flow)
B	A room containing a Grade A zone (the background environment for filling) and the area demarcated as the "Aseptic Filling Suite" (including final stage changing rooms)
C	Preparation of solutions to be filtered and production processing; component handling
D	Handling of components after washing; plasma stripping
U[a]	Freezers, computer conduits, store rooms, electrical cupboards, other rooms not in use, etc.

[a]U = unclassified. Unclassified areas are not monitored.

Table 16.2   **Comparison of EU GMP and ISO14644 cleanroom states at rest**

EU GMP	ISO 14644-1
A	5
B	5
C	7
D	8

"cleaner" the room class and the higher the number (such as "9") the room is considered to be "less clean." ISO 14644 class 5 is the critical zone where sterilized product, components, and product contact equipment are brought together and exposed.

The ISO classes and the EU GMP grades are approximately equivalent (although there are slight differences in the number of particles of a given size permitted). The table below compares the EU GMP grades and the ISO classes in the at rest state:

Cleanrooms have three different "states" of use. These are:

- As built;
- At rest (or static);
- In operation (or dynamic).

"As built" refers to the condition of a newly built cleanroom, with the operational qualification having been completed, at the point it is handed over to the user for performance qualification. There are two approaches to the construction of cleanrooms with regard to maintaining cleanliness. One method is to clean the facility at the end of the construction, often called "final super clean," while the other method, called "clean-build," requires continuous cleaning during construction. "Clean-build" protocols attempt to prevent contamination or capture contaminants at their source during the construction process. This clean construction protocol concept stresses the importance of exercising the discipline to build clean because it was widely believed that ultra-low levels of contamination cannot be reached using normal construction techniques and then by cleaning the facility afterwards.

For "at rest" conditions, there is a difference between European/ISO and US standards. The EU GMP defines the static state as a room without personnel present, following 15–20 min "clean up time," but with equipment operating normally. The US standards indicate that equipment is not running. "In operation" conditions are defined as rooms being used for normal processing activities with personnel present and equipment operating. Although both "static" and "dynamic" states are considered important by European regulators, for the United States the FDA tends only to focus on the dynamic state. This is because excursions within the dynamic state pose the bigger risk to product [7].

16.4   Isolators

Isolator technology is increasing being used in place of cleanrooms for critical activities such as aseptic filling and sterility testing. Isolators are part of a field called "barrier technology." Isolators are superior to cleanrooms in that the contamination risk is reduced through the construction of a barrier between the critical area (sometimes called the "microenvironment") and the outside environment.

Cleanroom and barrier isolator systems have four basic parts: the physical structure, the internal environment, the interaction technology, and the monitoring system. Thus, many of the test parameters described for cleanrooms will apply. It is also normal for isolators to be housed in cleanrooms (certainly those used for aseptic filling); this requires a further understanding of cleanroom disciplines.

Isolators will be positively or negatively pressured depending upon the required application. The main contamination risks from isolators are with the methods for transferring material in and out of the Isolator, especially as this requires personnel manipulations. These risks are minimized by the use of rapid transfer ports and by sanitizing isolators with a sterilant (vapor phase hydrogen peroxide is the most commonly used to decontamination cycle).

16.5   Cleanroom certification

Classification of critical cleanrooms is normally confirmed in the "in operation" state by taking nonviable particulate readings at a defined number of locations for 5.0 and 0.5 μm size particles (some pharmaceutical manufacturers opt to classify in the static state). The following frequencies (Table 16.3) are often adopted (as stated in ISO 14644-2):

The method of classification is based upon a number of locations in a cleanroom being monitored for particle counts (at the 0.5 μm size). Depending upon the grade of the room the sample size should either be one cubic meter per location or the total number of locations in the room should represent a sample size of one cubic meter. The number of room locations is the square root of the room in cubic meters and is defined in ISO14644-2.

16.6   Cleanroom testing

Once a room has been assigned a classification, certain environmental parameters (physical and microbiological) are to be met on a routine basis. For viable monitoring it is normal for the microbiologist to set action levels (and warning levels, which are equivalent to what are sometimes referred to as alert levels) based on an historical analysis of data.

The frequency of the assessment of other parameters (as described below) should be assessed based on a risk management approach. This approach should consider the room use and the risk to the product. Factors to consider may include [8]:

- room activities,
- exposure risk,
- room temperature

Table 16.3   **Cleanroom qualification frequencies**

Grade	Frequency of classification
A	Six-monthly
B	Six-monthly
C	Annually
D	Annually

- Process stage.
- Duration of process activities.
- Water exposure.

The emphasis should always be upon environmental control rather than simply environmental monitoring. That is, where a risk is identified, the risk should be minimized as part of a strategy of bringing the clean area into tighter control. Where a risk cannot be minimized but continues to exist then carefully targeted monitoring should be undertaken and the data reviewed and examined for trends, by a pharmaceutical microbiologist.

16.6.1   Physical parameters

There are a number of physical parameters which require examination on a regular basis. These parameters generally relate to the operation of HVAC systems and the associated air-handling chapters. Air handler, or air-handling chapter (AHU), relates to the blower, heating and cooling elements, filter racks or chamber, dampers, humidifier, and other central equipments in direct contact with the airflow. The key aspects which require testing are discussed below.

16.6.1.1   Air patterns and air movement

Airflows, for critical activities, need to be studied in order to show that air turbulence does not interfere with critical processes. There are two types of cleanrooms: turbulent flow or unidirectional flow, depending upon the required application. Unidirectional airflow areas are used for higher cleanliness states (such as aseptic filling), and they use far greater quantities of air than turbulent flow areas. Typically all critical rooms and zones within an aseptic filling area relating to batch filling should be assessed. Other critical processes may also be monitored. Airflows are studied using smoke generating devices and should be captured onto video tape or camera, and a report should be generated for each study. For air flow movement, air flow must be from a higher grade area to a lower grade area.

16.6.1.2   Airflows

Grade A zones (undirectional airflow devices in Grade B rooms) have a requirement for controlled air velocity and unidirectional air flow (either horizontal or vertical). These are monitored using an anemometer. The air velocity is designed to be sufficient to remove any relatively large particles before they settle onto surfaces [9].

This monitoring is performed routinely and during re-qualification exercises. The target range is 0.45 m/s (±20%).

16.6.1.3   Air changes

Each cleanroom grade has a set number of air changes per hour. A typical air conditioned office will have something between 2 and 10 air changes per hour in order to give a level of comfort. The number of required air changes in a cleanroom is typically much higher (at least 20, and somewhat higher than this in changing rooms). Air changes are provided in order to dilute any particles present to an acceptable concentration (thus, air change is a way of expressing the level of air dilution which is occurring). Any contamination produced in the cleanroom is theoretically removed within the required time appropriate to the room grade. Monitoring air changes is necessary because the re-circulation of filtered air is important for maintaining control of the clean area.

16.6.1.4 Clean up times

Connected to air changes is the time taken for a clean area to return to the static condition, appropriate to its grade, in terms of particulates. This is sometimes called the recovery rate. The target time for "cleaning up" is 15–20 min.

16.6.1.5 Positive pressure

Connected to the measurement of air flow is positive pressure. In order to maintain air quality in a cleanroom, the pressure of a given room must be greater relative to a room of a lower grade (ΔP). This is to ensure that air, and hence particulate contamination, does not pass from "dirtier" adjacent areas into the higher grade cleanroom (this can also be observed by smoke studies). Generally, this is 15–20 Pa, although some areas of the same grade will also have differential pressure requirements due to specific activities. The most commonly encountered problems relate to situations when cleanroom doors are opened, and here it can be difficult to maintain pressures.

Note: Pressure differentials (ΔP expressed in Pascals) are the relative pressures from a higher grade area into a lower one. These are guidance values taken from EU GMP Annex 1 Manufacture of sterile medicinal products.

16.6.1.6 HEPA filters

HEPA (High Efficiency Particulate Air) filters are used in cleanrooms in many different industries, including semiconductor, pharmaceutical medical devices, nuclear, and biotechnology. The main function of a HEPA filter is to provide clean air to the cleanroom. HEPA filters are replaceable, extended-media, dry-type filters in rigid frames with set particle collection efficiencies. HEPA filter is constructed with many pleated layers of filter media paper; this design prevents particles from freely passing through the filter as they become trapped and stick onto the filter fibers. The filters are designed to control the number of particles entering a clean area by filtration. In Grade A zones, HEPA filters also function to straighten the airflow as a part of the unidirectional flow. In order to measure the effectiveness of the filters, they are checked for leaks (a DOP test). Leakage is assessed by challenging the filters with a particle generating substance and measuring the efficiency of the filter.

Particles are mainly trapped (they stick to a fiber) by one of the following four mechanisms:

- Straining/sieving. This is defined as when a particle is too large and becomes trapped between two filter fibers.
- Interception, where particles following a line of flow in the air stream come within one radius of a fiber and adhere to it.

- Impaction, where larger particles are unable to avoid fibers by following the curving contours of the air stream and are forced to embed in one of them directly; this increases with diminishing fiber separation and higher air flow velocity.
- Diffusion, an enhancing mechanism is a result of the collision with gas molecules by the smallest particles, especially those below $0.1\,\mu m$ in diameter, which are thereby impeded and delayed in their path through the filter; this behavior is similar to Brownian motion and raises the probability that a particle will be stopped by either of the two mechanisms above; it becomes dominant at lower air flow velocities.

In Europe, HEPA filter integrity taken from BS EN 1822 Parts 1 and 2. For example, a filter with a 99.997% efficiency is based on the particle sizes $0.3\,\mu m$ and larger (i.e., theoretically only 3 out of 10,000 particles at $0.3\,\mu m$ size can penetrate the filter). This is typically assessed through thermally generated dioctylphthalate (DOP) (or specified alternative aerosol) particles, and a maximum clean-filter pressure drop test.

An ULPA filter (theoretically) can remove from the air at least 99.999% of dust, pollen, fungi, bacteria, and any airborne particles with a size of $0.12\,\mu m$ or larger.

16.6.1.7   Temperature, humidity, lighting, and room design

In terms of the general design of cleanrooms, they should:

- Be built of an airtight structure;
- The internal surfaces should be smooth and suitable for cleaning;
- The internal surface finish should be tough enough to resist chipping or powdering;
- The surfaces should be resistant to the cleaning agents used.

In addition to surfaces, parameters such as temperature and humidity should be controlled through the HVAC system. This is important for operator comfort and as a way of minimizing contamination (e.g., high temperatures which lead to excessive perspiration can reduce the efficiency of the cleanroom suit). HVAC systems use ducts (through ductwork systems) to deliver and remove air. Ducts deliver, most commonly as a part of the supply air, ventilation air. As such, air ducts are one method of ensuring acceptable indoor air quality as well as thermal comfort.

Grade B rooms should have set requirements for temperature and humidity ($18 \pm 3\,°C$ and $45 \pm 15\%$ relative humidity). These are monitored for operator comfort and to avoid a high temperature/humidity situation which may result in the shedding of microorganisms. Other clean areas have a temperature appropriate to the process step (e.g., if the process requires a cold room at $2–8\,°C$).

Lighting should be adequate, uniform and antiglare, to allow operators to perform process tasks effectively. A range of 400–750 lux is recommended.

Cleanrooms are specially designed rooms. The surfaces are constructed from materials that do not generate particles and are easy to clean.

16.7   Microbiological environmental monitoring

The areas of cleanroom monitoring which typically fall under the responsibility of the microbiology department are viable and nonviable methodologies. This forms the "environmental monitoring program." Environmental monitoring is a program which evaluates the cleanliness of the manufacturing or process environment; the effectiveness of cleaning and disinfection programs and the operational performance of environmental controls. Environmental monitoring was, arguably, an underdeveloped activity until the 1980s. Current programs are more sophisticated and are focused on contamination control rather than simply detecting contamination events [10].

The emphasis should always be upon environmental control rather than simply environmental monitoring. That is, where a risk is identified, the risk should be minimized as part of a strategy of bringing the clean area into tighter control. Where a risk cannot be minimized but continues to exist then carefully targeted monitoring should be undertaken and the data reviewed, and examined for trends, by a pharmaceutical microbiologist. Furthermore, part of environmental control involves designing facilities in an optimal way so that contamination is minimized, and where environmental monitoring becomes a tool in the assessment of control [11].

However, with no risk able to be reduced completely to zero, there remains an important role for environmental monitoring.

Nonviable monitoring is for air-borne particle counts. These are the same sizes of particles required for the classification (as described above): 0.5 and 5.0 μm. This is undertaken using an optical particle counter. Particle counters are used to determine the air quality by counting and sizing the number of particles in the air.

Viable monitoring is designed to detect levels of bacteria and fungi present in defined locations/areas during a particular stage in the activity of processing and filling a product. Viable monitoring is designed to detect mesophilic microorganisms in the aerobic state. However, some manufacturers may have requirements to examine for other types of microorganisms (such as anaerobes if nitrogen lines are used as a part of the manufacturing process).

Monitoring methods will all use either a general purpose culture medium like tryptone soya agar (TSA), which will be used at a dual incubation regime of 20–25 °C and 30–35 °C or two different culture media are used at two different temperatures, of which one of the media is selective for fungi (e.g., Sabouraud Dextrose agar, SDA). The choice of culture media, incubation times, and temperatures requires validating.

Viable microbiological monitoring is normally performed in the "in operation," as this represents a more realistic assessment of the challenge to the manufacturing process. Dynamic is interpreted as rooms being used for normal processing activities with personnel present.

There are different methods which are used for viable monitoring. These can be grouped into air and surface methods and into primary and secondary methods based on their theoretical efficiencies to recover micro-organisms (Table 16.4):

In addition to the classic methods outlined above, rapid microbiological methods are available. These include spectrophotometric particle counters. Rapid microbiological methods are considered in Chapter 17 [12].

16.7.1   Air sampling methods

For air monitoring, this is undertaken using agar settle plates (placed in the locations of greatest risk) or active (volumetric) air-samplers (to provide a quantitative assessment of the number of microorganisms in the air per volume of air sampled).

Table 16.4 **Microbiological viable monitoring methods**

	Air	Surface	Personnel
Method #1	Active air Sampler (cfu/m³)	Contact Plate (cfu/25 cm²)	Finger plate for Hands (cfu/5 fingers) Contact plate for gowns (cfu/25 cm²)
Method #2	Settle Plate (cfu/90 mm over "x" time period)	Swab (cfu/surface)	

16.7.1.1 Settle plates

A settle plate is an agar plate, placed in a defined location. The exposure time of the settle plate can be varied, although there is probably little value in exposing plates for less than 1 h. For consistency of sampling, for aseptic filling, the EU GMP Guide recommends a 4 h exposure time. This time should not be exceeded without strong justification, and even then there will probably be a challenge from the regulatory authority. For exposure times under 4 h, such as when a shorter activity is being monitored, the result obtained should be extrapolated using the simple equation:

$$\frac{Count}{Time\,exposed\,(min)} \times 240 = cfu\,/\,4\ h$$

The risk from any exposure is desiccation. The depth and condition of the agar are the key variables, as is the cleanroom environment. The agar in the plate will dry out faster if the airflow is excessively high or if the air humidity is low. Therefore, the exposure time of settle plates under the conditions of use.

16.7.1.2 Active air samples

Active (or volumetric or bioaerosol) air samplers are a slightly different measure of microorganisms in air than settle plates. As indicated earlier, the settle plate indicates the number of microorganisms that may deposit onto a surface; whereas, the active air-sampler indicates the number of microorganisms present in a given volume of air within the range of the air-sampler. Both of these approaches have merits and any comprehensive program will use both active air samples and settle plates.

Active air-samplers sample a defined quantity of air. The volume of air sampled is normally one cubic meter of air. This allows the data to be quantified as cfu/m³.

Active air-samplers generally fall into the following different models:

- slit to agar,
- membrane filtration,
- centrifugal samplers.

Where air samplers are required to be used in unidirectional airflow devices, the samplers should be isokinetic in operation as so not to disrupt the air stream [13].

16.7.2   Surface sampling methods

16.7.2.1   Contact plates

Surface contact plates are a common test for surface contamination. Contact plates are Petri-dishes filled with microbiological agar. The plate is filled to a level above the rim of the plate so that the agar surface extends upwards when dry. The plate has a typical diameter of 50–55 mm and a surface area of 25 cm^2.

The raised surface allows the agar to be pressed onto a surface. The design of the contact plate is, therefore, different to the standard Petri-dish, where the agar is contained within the Petri-dish.

The contact plate is a quantifiable method, because the contact between the plate and the surface provides a "mirror image" of the surface. Following incubation, this image transfer provides information relating to the number of microbial colonies and their relative position. The quantification is derived from the recording the number of colony forming units (cfu) per square centimeter. This act of replication gives the contact plate its alternative name: RODAC. This is an acronym for "Replicate Organism Detection and Counting," and the term is more commonly applied in North America.

In Grades A and B cleanrooms, in relation to aseptic processing, contact plates are taken from personnel hands (during processing activities) and from clothing. Personnel gown monitoring should be carried out at the end of a shift. This is because the sampling damages the fabric integrity of the suit through the moisture of the agar plate.

16.7.2.2   Swabs

Swabs are typically made up of sterile cotton tips, although swabs vary in the materials used for the applicator stick (either wood or plastic) and the tip (where materials like cotton, viscose, alginate, and so on are used). Some types of swabs require prewetting using a diluent (such as Ringer's solution, phosphate-buffered saline, sodium hexametaphosphate, sterile water, etc.) before use. Other types of swabs are contained with a transport medium. They are either contained within a transport medium or require prewetting with a suitable recovery medium. Swabbing is performed by rubbing a surface while rotating the swab—so that all parts of the tip are exposed—through a number of strokes (typically between 10 and 25).

Following the sampling, the swab can either be streaked out onto an agar plate (which is the least efficient method) or, where the appropriate tip has been used, dissolved in a diluent and tested either by pour plate or by membrane filtration (of which the latter is the most efficient method). The membrane filter will be either 0.45 or 0.22 µm, and this technique carries a further advantage in that disinfectant residues can be overcome through rinsing the filter.

16.7.3   Key aspects of the monitoring program

The microbiologist should establish the appropriate frequencies and durations for monitoring based on a risk assessment approach [14]. This applies to all cleanrooms where product is processed. In addition, in relation to other parts of this book, environmental monitoring should extend to the microbiology laboratory [15].

The sampling plan should take into account the cleanliness level required at each site to be sampled. It should consider which types of samples are appropriate that is air samples and/or surface samples. The choice of sample locations has to consider the nature of the work to be carried out in the production process or cleaning process and the impact that cleanroom operators and equipment (both fixed and portable) will have on the biocontamination levels. This requires a detailed understanding of how the cleanroom operates in terms of airflow velocity and direction and how these factors interact with the equipment and people in the cleanroom. Once this has been considered, the number of samples to be taken can be determined.

In terms of sampling frequency, in order to set initial limits for the microbial monitoring, a number of sampling operations are carried out. Initially the locations are sampled with the area at rest to give some base reference information. Thereafter the locations are sampled during the production activity. Assuming the product quality is satisfactory when the sampling was carried out then the monitoring data is used to set limits. Thereafter the test frequency can be reduced to a level that will still demonstrate control of biocontamination. The time of sampling should also take account of testing after a "cleandown," testing at the end of a shift, testing at times of the highest operator activity or high levels of materials in the area. In addition, testing after new installed equipment or routine maintenance work should be considered. If out of limits, microbial levels are found that it would also be standard to increase the sampling frequency until control is demonstrated after any corrective action has been taken.

The environmental monitoring plan should also include sample handling and incubation.

For microbiological monitoring parameters, Table 16.5 below lists those typically applied as maximal values (with appropriate warning and action levels set at some level below):

Where nonviable and viable monitoring levels are exceeded, it is typical to identify to species levels all the contaminants from Grades A and B areas and to have an understanding of the microflora from other areas. The microflora recovered should be compared with the microorganisms from other areas, and a comparison made.

The types of microorganisms can indicate the origin of contamination, The recovery of endospore-forming bacteria and fungi, for example, could indicate a problem with HEPA filters or insufficient pressure differentials between a clean and less clean area; the recovery of Gram-positive cocci is invariably an indication of personnel contamination; whereas the recovery of Gram-negative bacteria can signify water or dampness, or possibly fabric damage. The list below provides examples of some commonly occurring microorganisms and the various aspects of cleanroom operations that they are associated with [16]:

Airborne types
- *Bacillus* spp.
- *Micrococcus* spp.
- *Staphylococcus* spp.
- *Aspergillus* spp.

Table 16.5 Viable and non viable particulate requirements for cleanrooms

Parameters	Grade A		Grade B		Grade C		Grade D	
Nonviable	Particle size/m³		Particle size/m³		Particle size/m³		Particle size/m³	
Particulates Static state	3520 at 0.5 µm	20 at 5.0 µm	3520 at 0.5 µm	29 at 5.0 µm	352,000 at 0.5 µm	2900 at 5.0 µm	3,500,000 at 0.5 µm	29,000 at 5.0 µm
Particulates Dynamic state	3520 at 0.5 µm	20 at 5.0 µm	352,000 at 0.5 µm	2900 at 5.0 µm	3,520,000 at 0.5 µm	29,000 at 5.0 µm	Not defined	Not defined
Air samples (active) Dynamic state	cfu/m³ Action=1		cfu/m³ Action=10		cfu/m³ Action=100		cfu/m³ Action=200	
Air samples: Settle plates (passive) Dynamic state	cfu/event Action=1		cfu/event Action=5		cfu/event Action=50		cfu/event Action=100	
Surface samples at working height: Contact plates Dynamic state	cfu/25 cm² Action=1		cfu/25 cm² Action=5		cfu/25 cm² Action=25		cfu/25 cm² Action=50	
Surface samples at working height: Swabs Dynamic state	cfu/swab Action=1		cfu/swab Action=5		cfu/swab Action=25		cfu/swab Action=50	
Surface samples: Floor contact plates/swabs	cfu/device Action=1		cfu/device Action=10		Not defined		Not defined	
Drain swabs	N/A		N/A		Not defined		Not defined	
Finger plates Dynamic state	cfu/plate (hand) Action=1		cfu/plate (hand) Action=5		cfu/plate (hand) N/A		cfu/plate (hand) N/A	
Gowning (suit contact plate) Dynamic state	cfu/25 cm² Action=1		cfu/25 cm² Action=5		cfu/25 cm² N/A		cfu/25 cm² N/A	

- *Penicillin* spp.
- *Corynebacterium* spp.

Personnel contamination
- *Staphylococcus* spp.
- *Staphylococcus epidermidis*
- *Staphylococcus capitis*
- *Staphylococcus hominis*
- *Propionibacterium* spp.
- *Micrococcus* spp.
- *Trycophyton* spp.
- *Epidermophyton* spp.
- *Micosporon* spp.

Water/water sources
- Gram-negative rods in general
- *Pseudomonas* spp.
- *Alcaligenes* spp.
- *Stenotrophomonas* spp.
- *Burkholderia cepacia*
- *Ralstonia picketti*
- *Serratia* spp.
- *Flavobacterium* spp.

Such data can provide useful information in terms of contamination patterns. Areas from cross comparison may include:

- Cleanrooms or controlled environments,
- Associated manufacturing areas,
- Raw materials or components,
- Personnel, when appropriate,
- Utilities, e.g., water and compressed air,
- Finished product.

Action level excursions should be investigated using established OOS (out of specification or out of limits) procedures. The microbiologist should be at the forefront of such investigations.

Environmental monitoring is normally conducted by microbiologists. However, sampling is better carried out by process personnel in aseptic filling areas in order to minimize the number of personnel present within a cleanroom [17].

16.7.4 Personnel

Personnel working in controlled environments should be correctly trained in cleanroom disciplines and in aseptic technique. The biggest source of contamination in cleanrooms is people, and levels of contamination are the highest where training is poor (such as making rapid movements); gowning is not effective; or where an individual has a high propensity to "shed" skin resident microorganisms. Contamination shed by people includes a mix of viable and nonviable particulates. The precise pattern is unique to each individual.

Experiments have shown that personnel clothed in new, sterile cleanroom garments slough viable contamination at a rate of roughly one viable particulate to 10,000 nonviable particles [18]. During slow deliberate movements with the best possible clothing, operators will slough particulate and viable organisms. Therefore, the probability of human borne microbial contamination being released in the conventional cleanroom is one over the course of any reasonably long operational shift. The microorganisms typically associated with personnel are those residential or transient to skin such as the *Staphylococci* and the *Micrococci*.

The cleanroom disciplines which require control relate to clothing; entry and exit procedures and movements within the cleanroom. The US Food and Drug Administration (FDA) Centre for Drug Evaluation and Research (CDER) call for both initial and on-going training for cleanroom operators:

> *Appropriate training should be conducted before an individual is permitted to enter the aseptic processing area and perform operations. For example, such training should include aseptic technique, cleanroom behaviour, microbiology, hygiene, gowning, patient safety hazards posed by a non-sterile drug product, and the specific written procedures covering aseptic processing area operations. After initial training, personnel should be updated regularly by an on-going training program.*

Training should not only examine the practical aspects of cleanroom disciplines but manufacturing staff should also have an understanding of basic microbiology and how contamination can be spread in process areas. A feature of personal hygiene will also be important!

16.8 Aseptic technique

A key principle in cleanroom operation is aseptic technique. Aseptic technique is a procedure to minimize contamination, where the performance of the procedure is conducted in a manner that prevents the introduction of contamination.

Aseptic technique is a mental discipline as well as a physical activity. Personnel working in cleanrooms should consider all of the surfaces as "nonsterile" [19]. Within Grades A and B areas, disciplines become tighter. Here, for example, objects, including the operators' hands, must never be placed between the source of the air and a sterile object. The operators' hands and arms must always be kept at a level beneath that of open product containers. Sterile components should under no circumstances be touched directly with gloved hands; a sterilized tool should always used for this purpose. Since gloved hands and arms will enter the sterile field, they must never touch walls, floors, doors, or other surfaces. Strenuous lifting and moving of tanks, trolleys, and so on must not be done by operators assigned to work within or near the sterile field, because the more strenuous the activity, the higher the level of particle generation, and at least some of the particulate released by the operator will surely be viable microorganisms.

16.9   Other cleanroom disciplines

As a part of personnel training and cleanroom control, other cleanroom disciplines should be put over to staff. The most fundamental of these relate to movements. In higher grade areas, movements should be slow, controlled and deliberate as to minimize contamination. For example:

- No touching of the face with gloved hand. The glove is to be changed immediately when this occurs.
- No adjustment of hair is to be carried out in the cleanroom. All personnel are required to go to the changing room to tuck their hair under the hair net, after which the glove is to be replaced.
- No torn glove, booties or jumpsuit are to be worn in the cleanroom. They are to be changed immediately when they are found to be torn.
- No removing of items from underneath of the cleanroom jumpsuit.
- No running, fast walking, pushing and shoving, smoking, eating, or drinking in the cleanroom.
- No exposure of forearm bare skin. Jumpsuit and glove must be fully covered while working in the cleanroom.
- No scratching of head, combing of hairs.

Furthermore, disciplines can include such "dos" and "donts" as not allowing the following to be brought into or used in a cleanroom:

- Cosmetic make-up;
- Noncleanroom or regular papers;
- Noncleanroom compatible pen;
- Pencils, erasers, highlighter pen, and correction fluid;
- Photocopied cleanroom paper not contained in cleanroom compatible cover;
- Carton material items, wood, "stera-form," regular cloth, and regular paper materials;
- Rusty tools or equipment;
- When in doubt of other materials which is not stated on the above, please consult the supervisor.

16.9.1   Clothing

Airborne microorganisms, which are almost exclusively bacteria, are normally dispersed into the air around us from the surfaces of our skin cells. Cleanroom garments help to eliminate this source of contamination by acting as a "person filter" to prevent human particulate matter from entering the atmosphere of the cleanroom. Cleanroom clothing is made from fabrics that do not lint or disperse particles and act as a filter against particles dispersed from the person's skin and indoor, or factory, clothing. The type of clothing used in a cleanroom varies according to the type of cleanroom. In cleanrooms where contamination control is very important, personnel wear clothing that completely envelops them to ensure that particles and bacteria are not dispersed into the air such as a coverall, hood, facemask, knee-length boots, and gloves. In cleanrooms where contamination is not as critical, then a smock, cap, and shoe covers may be sufficient.

There are different gowning requirements for different grades of cleanroom. For example:

16.9.2   Grade A areas

Headgear to totally enclose hair and beard; face mask; no jewellery worn; sterilized gloves; protective sterile clothing consisting of one piece suit gathered at wrists with high neck and shedding virtually no particles or fibers and retaining particles shed by body; and trouser bottoms tucked into boots which cover captive shoes.

16.9.3   Grade C areas

Hair and beards covered; no jewellery worn (except for covered wedding rings); protective clothing consisting of one piece suit gathered at wrists with high neck and shedding virtually no particles or fibers; and captive shoes or overshoes worn.

16.10   Cleaning

Part of the control of a cleanroom and a way of minimizing contamination from personnel is established through cleaning procedures [20]. This is often an underplayed aspect of environmental control. Manufacturers spend large sums of money on installing HEPA filters and in establishing controlled HVAC and then proceed to neglect the basic aspects like regular and consistent cleaning. To eliminate potential contamination, all cleanroom surfaces such as rafters, interstitial spaces, duct work, plenum areas, ceiling panels, T-bars, lighting, ionizing grids, return air vents, walls, windows, work stations, equipment surfaces, cabinets, sinks, shelves, furniture, doors, pass-throughs, air locks, floor mats, floors and raised floors, and sub-fabric require periodic cleaning.

Cleaning in a pharmaceutical cleanroom is concerned about minimizing the number of particles and microorganisms (in relation to the grade of the cleanroom). It is important to establish which areas are to be kept clean and at what level. It can be useful to define the critical and general zones within the room and then work out what contaminants and levels are acceptable within these zones, what levels exist now. This can be demonstrated through environmental monitoring. This will give some idea on the frequency of cleaning needed, it maybe hourly, daily, weekly, monthly, and quarterly. Such data can allow the establishment of a documented cleaning program and schedule.

A further important consideration will be the selection of appropriate disinfectants and detergents. These will require validating and regular review against the microflora recovered.

16.11   Conclusion

This chapter has set out to explain the importance of environmental contamination control and has covered:

- Why contamination is a problem;
- Some contamination sources;

- Contamination control;
- Understanding cleanrooms;
- Some aspects of environmental monitoring.

In covering these topics, the chapter has emphasized that environmental, and hence contamination control is of great importance for all areas of manufacturing. This is not only to meet regulatory standards but also to reduce the risk to the product and patient. Here it is important to note that although the standards (and often the level of monitoring) are higher for Grades A and B areas, what happens in other areas contributes to the quality of adjacent areas. For example, if a change area of wash-bay has an operational problem or if staff do not follow changing or cleaning procedures then higher graded area can be put at risk. An appreciation of contamination sources and contamination control is required by all who work within the pharmaceutical sector.

The student will also appreciate that to achieve contamination control requires a multidisciplinary approach. Microbiologists must work with Engineers and Production staff in order to ensure that cleanrooms are constructed, maintained, and monitored to agreed standards that problems are addressed and that risks are evaluated.

References

[1] Sutton SVW. Pharmaceutical quality control microbiology: a guidebook to the basics. Bethesda, USA: PDA/DHI; 2007. p. 91–108.

[2] Halls N. Risk management: practicalities and problems in pharmaceutical manufacture. In: Halls N, editor. Pharmaceutical contamination control. Bethesda, USA: PDA/DHI; 2007. p. 171–204.

[3] Moldenhauer J. Personnel and their impact on cleanrooms. In: 3rd ed. In: Nema S, Ludwig JD, editors. Pharmaceutical dosage forms: parenteral medications, vol. 2. London: Informa Healthcare; 2010. p. 56–79.

[4] De Abreu C, Pinto T, Oliveira D. Environmental monitoring: a correlation study between viable and nonviable particles in clean rooms. PDA J Pharm Sci Technol 2004;58(1):45–53.

[5] Sandle T. Environmental monitoring. In: Saghee MR, Sandle T, Tidswell EC, editors. Microbiology and sterility assurance in pharmaceuticals and medical devices. New Delhi: Business Horizons; 2011. p. 293–326.

[6] Sandle T. Cleanroom design. In: Moldenhauer J, editor. Environmental monitoring: a comprehensive handbook, vol. 7. Arlington Heights, IL, USA: PDA/DHI; 2015. p. 3–28.

[7] Sandle T. Microbiological environmental monitoring in clean areas: using risk assessment. PMPS 2004;2004(Winter):105–7.

[8] Sandle T. Environmental monitoring risk assessment. J GXP Compliance 2006;10(2):54–73.

[9] Ljungqvist B, Reinmuller B. Some observations on environmental monitoring of cleanrooms. Eur J Parenteral Sci 1996;1996(1):9–13.

[10] Ackers J, Agallaco J. Environmental monitoring: myths and misapplications. PDA J Pharm Sci Technol 2001;55(3):176–84.

[11] Miele WH. The fundamentals of an environmental control program. In: 3rd ed. In: Nema S, Ludwig JD, editors. Pharmaceutical dosage forms: parenteral medications, vol. 2. London: Informa Healthcare; 2010. p. 80–90.

[12] Sandle T, Leavy C, Jindal H, Rhodes R. Application of rapid microbiological methods for the risk assessment of controlled biopharmaceutical environments. J Appl Microbiol 2014;116(6):1495–505.

[13] Agalloco J, Akers J. Sterile product manufacturing. In: Gad SC, editor. Pharmaceutical manufacturing handbook: products and processes. NJ, USA: Wiley; 2007. p. 99–136.

[14] Sandle T. Application of quality risk management to set viable environmental monitoring frequencies in biotechnology processing and support areas. PDA J Pharm Sci Technol 2012;66(6):560–79.

[15] Jimenez L. Environmental monitoring. In: Jimenez L, editor. Microbial contamination control in the pharmaceutical industry. New York: Marcel Dekker Inc.; 2004. p. 103–32.

[16] Sandle T. A review of cleanroom microflora: types, trends, and patterns. PDA J Pharm Sci Technol 2011;65(4):392–403.

[17] Halls N. Microbiological environmental monitoring. In: Halls N, editor. Microbiological contamination control in pharmaceutical clean rooms. Boca Raton, USA: CRC Press; 2004. p. 23–52.

[18] Agalloco J, Ackers J. Validation of manual aseptic processes. In: Agalloco J, Carleton F, editors. Validation of pharmaceutical processes. 3rd ed. Boca Raton: CRC Press; 2008. p. 333.

[19] Reich RR, Miller M, Patterson H. Developing a viable microbiological environmental monitoring program for nonsterile pharmaceutical operations. Pharm Technol 2003; 27(3): 92–100.

[20] Sandle T. Environmental monitoring: a practical approach. In: Moldenhauer J, editor. Environmental monitoring: a comprehensive handbook, vol. 6. River Grove, USA: PDA/DHI; 2012. p. 29–54.

Rapid microbiological methods   **17**

17.1   Introduction

Rapid microbiological method (RMM) technologies aim to provide more sensitive, accurate, precise, and reproducible test results when compared with conventional, growth-based methods. RMMs may be interpreted to include any microbiological technique or process that increases the speed or efficiency of isolating, culturing, or identifying microorganisms when compared with conventional methods.

Although the use of the word "rapid" is often used to describe the range of techniques employed, some of the methods included here do not give a more rapid result but instead a more accurate, precise, or detailed result, so providing more information on which to base a decision. Alternative methods are sometimes referred to in relation to rapid methods, although alternative methods need not necessarily be "rapid." Where a compendia test is described (such as the sterility test), it is assumed to be the standard method, and from this, an alternative method is simply one that is different from the method described in the pharmacopoeia.

Rapid methods normally involve some form of automation, and the methods often capture data electronically. With several different technologies available on the marketplace, the microbiologist has a difficult, and sometimes expensive, choice to make in selecting the optimal method. When considering rapid methods, the new method must offer a higher level of quality assurance. There needs to be a clear and demonstrable benefit in adopting the alternative method. Several advantages are discussed in this chapter.

This chapter, while addressing some of the emerging technologies, is not so much about the different RMMs that are available; it is more concerned with the considerations that are needed for their selection. As such, the chapter provides some advice for the microbiologist to consider when drawing up a rationale for the selection of a rapid or alternative microbiological method. The chapter does describe different technologies, but it does not describe different commercial companies (to do so could date the chapter and could risk giving preference to a system that is not necessarily the most robust). The reader with an interest in a given technology and application should consult contemporaneous literature.

17.2   Changing world of microbiology

Conventional microbiological methods, including those long-established and described in the European (Ph. Eur.), Japanese, and US Pharmacopeia (USP), have served microbiologists well over the past century and have helped to ensure the production of microbiologically safe products. For example, a wide range of microbiological methods has been successfully verified using plate count methods to enumerate and identify

Pharmaceutical Microbiology. http://dx.doi.org/10.1016/B978-0-08-100022-9.00017-7

microorganisms (within an accepted margin of error [1]). However, conventional methods have limitations. These limitations include the time taken to produce a result and the inability of many methods to recover all of the microorganisms that might be present in a sample.

Considering these issues further, the time taken to produce a result relates to the incubation period required for conventional methods, which rely on agar as a growth medium, or for microorganisms to grow in broth culture. Such methods are relatively slow, and results are only available after an incubation period (somewhere between 2 and 10 days, depending on the application) [2].

A further limitation is culturability and the issue of viable but nonculturable (VBNC) microorganisms. Many bacteria, although maintaining metabolic activity, are nonculturable due to their physiology, fastidiousness, or mechanisms for adaptation to the environment. Some research suggests, for example, that less than 10% of bacteria found in cleanrooms are culturable [3]. Thus, it stands that some RMMs, especially those that do not rely on growth, may provide a higher recovery count as compared with traditional methods. Some rapid methods produce results where the number or types of microorganisms can be measured. With rapid methods that do not directly "grow" microorganisms, such as those that detect metabolic activity, it is possible to correlate the new measurements, such as a fluorescing unit, with the old measurement (i.e., the colony-forming unit; CFU) and establish new acceptance levels.

These concerns with limitations of conventional methods, as well as the possibilities afforded by technological advances, led to an emerging new generation of rapid and alternative microbiological methods. RMMs and alternative microbiological methods include any microbiological technique or process that increases the speed or efficiency of isolating, culturing, or identifying microorganisms when compared with conventional methods [4].

As to what rapid methods are, according to the US Food and Drug Administration (FDA) [5]:

> *RMMs are based on technologies which can be growth-based, viability-based, or surrogate-based cellular markers for a microorganism (i.e., nucleic acid-based, fatty acid-based). RMMs are frequently automated, and many have been utilized in clinical laboratories to detect viable microorganisms in patient specimens. These methods reportedly possess increased sensitivity in detecting changes in the sample matrix (e.g., by-products of microbial metabolism), under conditions that favour the growth of microorganisms.*

Although the use of the word "rapid" is often used to describe the range of techniques employed, some of the methods included within this collective do not give a more rapid result; they instead provide a more accurate, precise, or detailed result (and thus the term "alternative" is employed).

RMMs can be applied to a range of microbiological tests, including raw materials, water, intermediate products, final products, and environmental monitoring. There is a sufficient range of RMMs to provide an assessment of the microbiological quality throughout an entire production operation. RMMs may also be used by research and development. For example, in understanding formulations of products better in terms of whether microorganisms are likely to survive or be killed.

RMMs are essentially used as alternatives to four major types of conventional microbiological determinations [6–8]:

- qualitative tests for the presence or absence of microorganisms (e.g., enrichment turbidity measurements of growth). For example, to determine if *Escherichia coli* is in a sample of water;
- quantitative tests for enumeration of microorganisms (e.g., plate count methods to determine the bioburden of a sample);
- quantitative tests for potency or toxicity (e.g., what level of endotoxin is in the sample?);
- identification tests (e.g., biochemical and morphological characterization).

17.3   Advantages of rapid methods

Looking at some of the advantages afforded by rapid methods further, aside from the time-to-result, another important area is throughput. Most rapid systems allow for higher volumes than the traditional method. In environments with considerable volumes of raw ingredients, in-process batches and final products to test, a high throughput can confer an important advantage for maintaining manufacturing up-time and moving an inventory as quickly as possible.

Furthermore, RMMs can assist with:

- designing more robust processes that could reduce the opportunities for contamination (fitting in with some quality-by-design objectives);
- developing a more efficient corrective and preventive action process;
- confirming that the process is in a continuous state of microbiological control through "real-time" monitoring (that meets some process analytical test objectives);
- assisting with continuous process and product improvement.

Other advantages include labor efficiency and error reduction. Reducing errors is one of the greatest potential benefits of rapid enumeration. While some methods require extra human intervention and, thus, create greater potential for mistakes, others automate the most error-prone processes. Microbial counting, incubation changeovers, and data entry can all become far more reliable given the right equipment.

Arguably, RMMs enable a proactive approach to be taken to instances of microbial contamination, especially in relation to out-of-specification results. Here, RMMs enable quicker responses to out-of-trend situations through providing real-time or near real-time results. This allows corrective actions to be taken earlier.

Furthermore, when considering an RMM, the new method must offer a higher level of quality assurance. There needs to be a clear and demonstrable benefit in adopting the alternative method. Examples of this include:

- the ability to make critical business decisions more quickly;
- the prevention of recalls through greater method sensitivity to microorganisms;
- the detection of "objectionable" microorganisms;
- recovery of higher or more accurate microbial numbers;
- potential reduced stock holding through faster release times;
- improvement in manufacturing efficiency;
- a more proactive, rather than reactive, decision making.

It is because of these advantages that RMMs are areas of considerable investment by vendors and attract interest from microbiologists.

17.4   Regulatory acceptance

RMMs are accepted by the major global regulatory agencies. For example, in 2011, the FDA published their new strategic plan entitled *Advancing Regulatory Science at FDA* [9]. In Section 17.3, the FDA seeks to "support new approaches to improve product manufacturing and quality." With regard to control and reduction of microbial contamination in products, the FDA supports those who:

- develop sensitive, high-throughput methods for the detection, identification, and enumeration of microbial contaminants and validate their utility in assessing product sterility;
- develop and evaluate methods for microbial inactivation/removal from pharmaceutical products that are not amenable to conventional methods of sterilization;
- evaluate the impact of specific manufacturing processes on microbial contamination;
- develop reference materials for use by industry and academia to evaluate and validate novel methods for detecting microbial contamination.

17.5   Types of rapid microbiological methods

RMM or alternative methods can be categorized into multiple means. One way is based on the technology or application. Here, based on the Ph. Eur., the methods can be grouped into six categories.

17.5.1   Growth-based methods

Growth-based methods are those where a detectable signal is usually achieved following a period of subculture (e.g., electrochemical methods). These methods generally involve the measurement of biochemical or physiological parameters that reflect the growth of microorganisms. These methods aim to decrease the time at which one can detect actively growing microorganisms. The methods continue to use conventional liquid or agar media. In summary, they include:

- Impedance microbiology (measurable electrical threshold during microbial growth). With these methods, the relationship between capacitance at the electrode surface and conductance from ionic changes in the media from the by-products produced during bacterial growth allows for the calculation of impedance. With this, increases in capacitance and conductance result in decreased impedance, which is indicative of bacterial growth;
- The detection of carbon dioxide. Such methods deploy an internal colorimetric carbon dioxide sensor. The sensor can be placed inside a media bottle (separated from the media by a semipermeable membrane). As carbon dioxide is produced by microbial metabolism diffuses across the membrane, it dissolves in water in the sensor and generates hydrogen ions, which result in a color change detected by a colorimetric detector. Light emitted by

the detector reflects off the sensor onto a photometer. The resulting voltage signal is proportional to the intensity of the reflected light and to the concentration of carbon dioxide in the media bottle. As an alternative, headspace pressure monitors work in a similar way. Here, headspace pressure platforms detect growth as a result of consumption or production of gases in the headspace of sealed media bottles causing conformational changes in the geometry of the septum;
- The utilization of biochemical and carbohydrate substrates. In a similar way, detectors present in media can detect microbial reactions;
- The use of digital imaging and auto-fluorescence for the rapid detection and counting of microcolonies. This method overcomes the problem with culture methods where the visualization of colonies requires several days. The rapid method uses lasers to scan microorganisms growing on membrane filters with optical imaging using a digital camera. This technique allows detection and enumeration of microcolonies within a few days. An advantage of this technique is that the microorganisms remain viable for identification after colonies become visible;
- Fluorescent staining and enumeration of microcolonies by laser excitation. This method is similar to the one described above;
- Selective media for the rapid detection of specific microorganisms. This standard approach can be automated through the detection of specific nucleic acid sequences Such systems break down samples at the genetic level, using a polymerase chain reaction (PCR) to detect bacteria and other microbes.

17.5.2   Direct measurement

Direct measurement is where individual cells are differentiated and visualized (e.g., flow cytometry). These methods generally use viability stains and laser excitation for the detection and quantification of microorganisms without the need for cellular growth. These methods include:

- demonstration of direct labeling of individual cells with viability stains or fluorescent markers with no requirement for cellular growth;
- flow cytometry (individual particles are counted as they pass through a laser beam) and solid-phase cytometry (staining and laser excitation method). With these technologies, the ability to stain microorganisms with dyes such as propidium iodide, which is impermeable to cells with intact membranes, and thiozole orange, which is permeable to all cells, allows a differentiation of viable and nonviable bacterial cells in fluid media.

17.5.3   Cell component analysis

Cell component analysis is where the expression of specific cell components offers an indirect measure of microbial presence (e.g., genotypic methods). These methods generally involve the detection and analysis of specific portions of the microbial cell, including adenosine triphosphate (ATP), endotoxin, proteins, and surface macromolecules. The methods include:

- ATP bioluminescence (the generation of light by a biological process). With this method, the addition of a substrate to a membrane surface yields fluorescence following exposure to microbial ATP. This is because ATP is the main chemical energy source of all living cells.

Detection systems based on the bioluminescence exploit the chemical release of ATP from microorganisms. ATP reacts with luciferase and a photon counting imaging tube detects photons released by this reaction. A computer monitor then represents the photons detected. There are both qualitative and quantitative systems available;

- Endotoxin testing (*Limulus* amebocyte lysate; LAL). As Chapter 11 describes, automated LAL testing can provide results within minutes regarding the presence of bacterial endotoxin in a sample;
- Fatty acid analysis (methods that utilize fatty acid profiles to provide a fingerprint for microorganism identification). This method is discussed in Chapter 9;
- Matrix-assisted laser desorption ionization—time of flight (MALDI-TOF) mass spectrometry (microbial identification). This method is discussed in Chapter 9.

17.5.4   Optical spectroscopy

Optical spectroscopy methods utilize light scattering and other optical techniques to detect, enumerate, and identify microorganisms (e.g., "real time" airborne particle counters). These methods include:

- Real-time and continuous detection, sizing, and enumeration of airborne microorganisms and total particles. These methods are applied to the monitoring of cleanrooms. With this technology, air is drawn into an instrument. Particles that pass through a 405-nm diode laser are sized and enumerated using a Mie scattering particle counter. At the same time, particles that contain biological targets, such as nicotinamide adenine dinucleotide (NADH), riboflavin, and dipicolinic acid, will auto-fluorescence as they pass through the laser, and a separate fluorescence detector will record these as biological events.

17.5.5   Nucleic acid amplification

Nucleic acid amplification technologies are those such as PCR-DNA amplification, ribonucleic acid (RNA)-based reverse-transcriptase amplification, 16S rRNA typing, gene sequencing, and other novel techniques. These methods include:

- riboprinting: 16S sequence of rRNA is highly conserved at the genus and species level. This method is discussed in Chapter 9;
- PCR methods for targeting specific microorganisms (millions of copies of the target DNA in a short period of time). This method is discussed in Chapter 9;
- gene sequencing (specific dye labeling). This is variant of the PCR method. With the method, a genetic analyzer separates the fragments by size and a laser detects the fluorescence color from each dye, producing a full gene sequence of the target DNA.

17.5.6   Microelectrical–mechanical systems

Microelectrical–mechanical systems (MEMS) utilize microarrays, biosensors, and nanotechnology to provide miniaturized technology platforms. These methods include:
- Microarrays (DNA chips) so-called "lab-on-a-chip technology," evolved from Southern Blot technology, to measure gene expression (e.g., mycoplasma detection). With this method, DNA is extracted, and PCR performed using primers specific for conserved and species-specific regions the microbial genome. The fluorescently labeled fragments are then hybridized to the microarray chip. The chip contains probes for species-specific targets.

17.6 Selection of rapid microbiological methods

It is important that care must be taken in choosing an RMM or alternative method for a particular application. The method must determine a product's critical quality attribute and adhere to appropriate good manufacturing practice principles and validation requirements [10].

In some ways, the process of introducing an RMM or alternative method does not differ significantly when compared with implementing a conventional method. The key points are ensuring the method is validated and shows acceptable recovery rates or accurate identification does not differ whether rapid or conventional methods are used [11].

When choosing to implement an RMM, it is important to ensure the new method is appropriate for the company's formulations, facilities, and personnel. For example, the introduction of a method with a higher level of sensitivity needs to be aligned with the existing bioburden in raw materials, environment, and finished products.

Guidance for the implementation of rapid methods is available from both the USP and the Ph. Eur.:

- USP <1223>, *Validation of Alternative Microbiological Methods* [12];
- Ph. Eur. 5.1.6., *Alternative Methods for Control of Microbiological Quality* [13];
- Ph. Eur. 2.6.27, *Microbiological Control of Cellular Products* [14].

In addition, the FDA did issue a draft *Guidance for Industry Validation of Growth-Based Rapid Microbiological Methods for Sterility Testing of Cellular and Gene Therapy Products*, defining validation criteria for growth-based rapid or alternative microbiological methods. The document never became a final text, and it was withdrawn in 2015. From an industry perspective, the Parenteral Drug Association (PDA) has published a useful guide for implementation [15].

There are several considerations to be made and steps to be taken for the implementation of RMMs. These are discussed below.

17.6.1 Key considerations

An important consideration is to decide what is wanted from a rapid method and to consider this alongside a cost–benefit analysis. The first step is to consider the following questions:

- what do I want to achieve?
- what budget do I have?
- what technologies are available?
- what technologies are 'mature'? Who else is using them?
- how "rapid" is the rapid method?
- what papers have been published on the subject? Are these "independent"?
- what have regulators said?

The above can form part of a risk–benefit consideration. Risk–benefit analysis should focus on [16]:

- the defined purpose for the test method;

- the type and depth of information required;
- the limitations of the conventional method and what the rapid method might be able to offer.

Next, a more detailed assessment should be undertaken. This includes considering such factors as time, accuracy, and automation.

With time, factors to consider are:

- time taken to prepare the test; is the rapid method faster, equivalent, or slower?
- time taken to conduct the test;
- sample throughput;
- time to result;
- whether there is a reduction in the time taken to conduct complimentary tests;
- whether more or less time is required for data analysis;
- whether results-reporting is simplified or more efficient?

With accuracy, issues to consider include:

- if the rapid method will lead to a reduction in human error;
- if there is a reduction in subjectivity;
- whether the alternative method will detect more accurately in comparison to a conventional method?
- whether there is a need for the rapid method to detect what a cultural method cannot?

Other considerations include:

- if there is a need for the electronic capture of data?
- whether the method needs to be automated?
- if there is a need for connecting apparatus or linking the method to a laboratory information management system (LIMS)?

17.6.2 Internal company obstacles

The conventional microbiological methods currently used are, generally, already approved and provide meaningful data. Consequently, there may be reluctance within companies to change procedures and adopt RMMs. Thus, arguments relating to the benefits of implementing RMMs may need to be explored.

Furthermore, there may be reluctance to adopt RMMs because of the capital investment in equipment, training, and possible adaption of current manufacturing processes as well as the time and cost of the important validation required before use. The financial implications are naturally important considerations, and it is recommended that discussions on whether or not to employ new RMMs should involve multidisciplinary personnel (e.g., senior management, quality unit, microbiology, production, business development, finance, and members of supply chain). With business issues, one of the key concerns is return-on-investment. This can be assessed by considering the following:

- operating costs of the conventional method;
- operating and investment costs of the RMM;
- cost-savings and cost avoidances of the rapid method.

The following questions can help with this step:

- how much will the validation cost?
- how long will the validation take?
- how many personnel will the validation require?
- how many tests will be needed to run for the validation?
- does the validation require a comparison with another (existing) method?
- how will the data be analyzed and reported?

The cost of implementation should not be considered in isolation; the cost–benefit to the business in terms of higher quality assurance, reduced stock inventory, and quicker release of product may generate cost reduction to the business in excess of the cost of implementation. Capital outlay and running costs will depend upon the RMM chosen, and the equipment purchased.

Other aspects that can support a business case include:

- online/at-line systems can result in reduced microbiology testing and finished product release cycle times;
- RMMs can assist in more immediate decisions on in-process material;
- reduced repeat testing and investigations;
- maximized warehousing efficiencies by way of reduced inventory holding;
- reduction in plant downtime/return from shut downs.
- increased production yield—shift to continuous manufacturing;
- maximized analyst output by eliminating waste activity.

17.6.3 Validation

When choosing an RMM, consideration should be given to how it is going to be validated. Any methods that are being adopted need to yield results equivalent to or better than the method currently used that already gives an acceptable level of assurance. In addition, the RMM and the method currently used should be run in parallel for a designated time as a condition of approval.

Validation will be cantered on two key aspects: the assessment of the equipment and an assessment of the materials that the RMM will assess to demonstrate that microorganisms can be recovered from the material under test [17].

The validation strategy should reflect the RMM selected. Some methods that are based on the analytical chemistry will suit validation criteria that include accuracy and precision, specificity, limit of detection, limit of quantification, linearity and range, and ruggedness and robustness. However, microbiology methods do not necessarily lend themselves to this approach to validation (in that not all of these criteria will be applicable), as FDA indicates [18]:

> While it is important for each validation parameter to be addressed, it may not be necessary for the user to do all of the work themselves. For some validation parameters, it is much easier for the RMM vendor to perform the validation experiments.

Therefore, the following validation strategy is recommended:

- define the characteristic of the current test that the RMM is to replace;

- determine the relevant measures that establish equivalence of the RMM to the current method. This may require statistical analysis;
- demonstrate the equivalence of the RMM to the current method in the absence of the product sample;
- demonstrate the equivalence of the RMM to the established method in the presence of the test sample.

More specifically, with certain groups of methods, these various validation considerations can be interpreted as:

(a) qualitative methods
  - accuracy and precision, a presence absence test = low number of positives of a low microbial count (fewer than 10 CFU);
  - specificity = growth promotion test;
  - limit of detection = inoculate at fewer than five CFU in both the pharmacopoeia method and the RMM to be tested over several replicates;
  - robustness = different variations of the normal test conditions (e.g., different analysts, different instruments, and different reagent lots).

(b) quantitative methods
  - accuracy = suspensions at the upper end of the expected range and then serially diluted down and testing alongside the compendial method. The level of agreement should not be less than 70% compared with the compendia test;
  - precision = a statistically significant number of replicates should be used. The level of variance should generally be within the 10–15% and should not be larger than that found within the pharmacopoeia method;
  - specificity = carried out using a range of microorganisms;
  - limit of quantification = the lowest number of microorganisms that can be reliably counted;
  - linearity = a directly proportional relationship between the concentration of microorganisms used and those expressed in the RMM;
  - range = the results found in precision, accuracy, and linearity can be used here in order to determine the upper and lower limits of detection of the RMM;
  - robustness = different variations of the normal test conditions (e.g., different analysts, instruments, and reagent lots).

During the course of validation, deviations from the established criteria may occur. The implications of these will depend upon the seriousness of the issue and the degree of drift from established parameters. The deviation may or may not lead to a recommencing of the validation after an appropriate change has been made. In the most serious cases, the deviation can lead to the abandonment of the qualification and the rejection of the equipment or system. All deviations require a deviation report to be generated. Deviation reports must be reviewed by a competent expert and be accepted by quality assurance.

With the equipment qualification aspect, validation normally begins with the validation plan (VP). The VP is a document that describes how and when the validation program will be executed in a facility. The VP document will cover some or all of the following subjects:

- introduction;

- plan origin and approval;
- derivation;
- scope of validation activities;
- validation objectives;
- VP review;
- roles and responsibilities;
- an overview of activities;
- division of responsibilities;
- system description;
- overview of system;
- overview of process;
- system description;
- validation approach;
- site activities;
- documentation and procedures;
- scope of documentation;
- validation schedule of activities;
- project master schedule;
- references.

From this plan, equipment validation is normally achieved through appropriate installation qualification (IQ), operational qualification (OQ), and performance qualification (PQ) [19]. Here:

- IQ provides documented evidence that the equipment has been provided and installed in accordance with its specification. The IQ demonstrates that the process or equipment meets all specifications, is installed correctly, and all required components and documentation needed for continued operation are installed and in place;
- OQ provides documented evidence that the installed equipment operates within predetermined limits when used in accordance with its operational procedures;
- PQ provides documented evidence that the equipment, as installed and operated in accordance with operational procedures, consistently performs in accordance with predetermined criteria and thereby yields correct results for the method.

17.6.4   Method transfer

If a validated method is transferred to another laboratory (including third parties), then appropriate change management should be in place. Full validation of the equipment (IQ, OQ, PQ) will need to be carried out. Full validation of the method may not be required, but, as a minimum, it needs to be demonstrated that the method gives equivalent or comparable results to the original laboratory. Any changes to formulations need to be assessed to determine if full or partial revalidation of the method is required [20].

17.6.5   Training

It is important that when RMMs are introduced, sufficient training is provided to ensure a successful and complete implementation of the new methods. This should include the microbiologists and other personnel involved in the running of the tests and should also take account of the laboratory or manufacturing facilities. Different rapid methods may also require different steps for sample preparation. Rapid methods that require different preparation steps than traditional methods will require additional training and standard operating procedure (SOP) updates.

Qualified microbiologists will still be required to interpret and manage the data, continue to develop the method, and ensure that correct decisions are made. This should form part of the overall microbial quality management system. Consideration can also be given to how rapid methods can be used within the laboratory or within the facility (such as tools for risk assessment) [21].

17.6.6   Expectations from the vendor

Outside of the suitability of the technology, there are a number of points that need to be satisfied in considering a specific technology, most notably the experience of the vendor itself. The following points can be useful:

- what is the vendor's expertise to date?
- is the vendor in a position to support your validation process?
- does the vendor have the relevant quality management system procedures in place?
- what stage is the vendor at in terms of development? For example, is the company financially sound?
- is the technology known to regulators?
- has the vendor made any product filings to regulators?
- does the vendor supply relevant documentation with the technology? For example, design of documents, providing material standards, and so forth.
- does the vendor provide training to analysts?
- is the vendor in a position to react with a reasonable response time to technical issues?
- how often does the vendor envisage system/software updates, and how will these be handled?

17.7   Summary

This chapter has outlined some of the key considerations to be made when deciding whether to adopt a rapid method and the subsequent selection between the different types of rapid methods that are available. The chapter has not set out to differentiate between different technologies (this itself is a rapidly developing field) but more to offer general advice to those tasked with making the selection and undertaking the work required to qualify the method so that it is available for the laboratory or process area to use.

References

[1] Sutton S. Accuracy of plate counts. J Validation Technol 2011;17(3):42–6.
[2] Gray JC, Staerk A, Berchtold M, Hecker W, Neuhaus G, Wirth A. Growth promoting properties of different solid nutrient media evaluated with stressed and unstressed

micro-organisms: prestudy for the validation of a rapid sterility test. PDA J Pharm Sci Technol 2010;64:249–63.

[3] Sandle T. A review of cleanroom microflora: types, trends, and patterns. PDA J Pharm Sci Technol 2011;65(4):392–403.

[4] Duguid J, Balkovic E, du Moulin GC. Rapid microbiological methods: where are they now? Am Pharm Rev 2011;9:10–18. http://www.americanpharmaceuticalreview.com/ Featured-Articles/37220-Rapid-Microbiological-Methods-Where-Are-They-Now.

[5] FDA. Guidance for industry validation of growth-based rapid microbiological methods for sterility testing of cellular and gene therapy products. Bethesda, MD: US Food and Drug Administration; 2008.

[6] Moldenhauer J. Overview of rapid microbiological methods. In: Elwary S, Turner A, Zourob M, editors. Principles of bacterial detection: biosensors, recognition receptors and microsystems. New York: Springer; 2008. p. 49–79.

[7] Miller MJ. Encyclopedia of rapid microbiological methods. Bethesda, MD: Parenteral Drug Association and Davis, Healthcare International Publishing, LLC; 2005, p. 103–35.

[8] Noble RT, Weisberg SB. A review of technologies for rapid detection of bacteria. J Water Health 2005;3:381–91.

[9] FDA. Strategic plan for regulatory science. Bethesda, MD: US Food and Drug Administration; 2011. http://www.fda.gov/downloads/ScienceResearch/SpecialTopics/ RegulatoryScience/UCM268225.pdf.

[10] Denoya C, Colgan S, du Moulin GC. Alternative microbiological methods in the pharmaceutical industry: the need for a new microbiology curriculum. Am Pharm Rev 2006;9:10–8.

[11] Griffiths MW. Rapid microbiological methods with hazard analysis critical control point. J AOAC Int 1997;80(6):1143–50.

[12] USP. Validation of alternative microbiological methods. In: United States Pharmacopoeia. 34th ed. Rockville, MD: The United States Pharmacopeial Convention; 2011.

[13] European Pharmacopeia. Alternative methods for control of microbiological quality. 7th ed. Strasbourg, FR: European Directorate for the Quality of Medicines; 2011 [Section 5. 1. 6].

[14] European Pharmacopoeia. Microbiological control of cellular products. 7th ed. Strasbourg, FR: European Directorate for the Quality of Medicines; 2011 [Section 2. 6. 27].

[15] PDA evaluation, validation and implementation of alternative and rapid microbiological methods. Technical report no. 33 (2nd revision). Bethesda, MD, USA: Parenteral Drug Association; 2013.

[16] Cundell AM. Opportunities for rapid microbial methods. Eur Pharm Rev 2006;1:64–70.

[17] Sutton S. Validation of alternative microbiology methods for product testing quantitative and qualitative assays. Pharm Technol 2005;29:118–22.

[18] Riley B. A regulators view of rapid microbiology methods. Eur Pharm Rev 2011;16(5):3–5.

[19] Miller M. The implementation of rapid microbiological methods. Eur Pharm Rev 2010;15(2):24–6.

[20] Jimenez L. Rapid methods for pharmaceutical analysis. In: Jimmenez L, editor. Microbiological contamination control in the pharmaceutical industry. New York: Marcel-Dekker; 2004. p. 147–82.

[21] Sandle T, Leavy C, Jindal H, Rhodes R. Application of rapid microbiological methods for the risk assessment of controlled biopharmaceutical environments. J Appl Microbiol 2014;116(6):1495–505.

Risk assessment and microbiology

18

18.1 Introduction

Risk assessment and risk management are important factors in the manufacture and quality control of pharmaceuticals and biopharmaceuticals. Although pharmaceutical manufacturers can rely upon process validation and on in-process and finished product testing to assure the quality of the drug products reaching the patient, it is arguably more effective for pharmaceutical manufacturers and regulators to focus on product and process defects that have significant impact on the patient. This philosophical point has formed the basis of the risk assessment and analysis initiatives that have been implemented globally within the pharmaceutical industry, to different degrees of application.

Risk is always prevalent with the production of pharmaceutical products, and in relation to microbiological risks in particular. The manufacturing and use of a drug product, including its components, necessarily entail some degree of risk. The risk to its quality is just one component of the overall risk. It is important to understand that product quality should be maintained throughout the product lifecycle such that the attributes that are important to the quality of the drug product remain consistent with those used in the clinical studies. While risks have been ever-present, the formal review of risk became part of the regulatory landscape at the turn of the twenty-first century.

The drive toward risk assessment began when the US Food and Drug Administration (FDA) announced what was then a new initiative: "Pharmaceutical cGMP (current good manufacturing practice) for the 21st century in 2002." This resulted in the FDA implementing a new science-based regulatory strategy emphasizing quality systems, risk assessment, and risk management. Several guidance documents have been issued to support the initiative. These included the adoption of a quality system approach to pharmaceutical the organization of good manufacturing practice (which also became a model for conducting inspections); recommendations for the use of process analytical technology (PAT) to allow for "real-time" process monitoring; and the use of a range of risk assessment tools and techniques [1].

Around the same time, the International Conference on Harmonization (ICH) published a document (ICH Q9) on Quality Risk Management. This was followed by companion documents outlining pharmaceutical development (Q8) and quality systems (Q10). Of these, the document of the greatest importance was ICH Q9. This document outlined a risk management strategy that involved the concepts of risk identification, assessment, control, communication, and review. In time, ICH Q9 became "adopted" by FDA and incorporated into EU GMP.

Pharmaceutical Microbiology. http://dx.doi.org/10.1016/B978-0-08-100022-9.00018-9

What the recent regulatory approaches do is direct the manufacturer to the fact that an effective quality risk management approach can further ensure the high quality of the medicinal product to the patient by providing a proactive means to identify and control potential quality issues during development and manufacturing. Additionally, use of quality risk management can improve the decision making if a quality problem arises. Effective quality risk management can facilitate better, and more informed decisions, can provide regulators with greater assurance of a company's ability to deal with potential risks and can beneficially affect the extent and level of direct regulatory oversight.

This chapter considers the nature of risk, outlines some risk assessment approaches, and contextualizes these within the context of microbiological risks.

18.2 The nature of risk

Risk is a difficult concept to define, in that it can only be considered as a relationship to something else (that is comparing one risk to another provides some kind of measure as to whether the risk is greater or lesser than the other). In more systematic terms, risk can be defined as the combination of the probability of occurrence of harm and the severity of that harm [2]. Thus, risk can be derived from a consideration of:

1. what might go wrong?
2. what is the likelihood (probability) it will go wrong?
3. what are the consequences (severity)?

Failure modes and effects' analysis (FMEA), as a risk analysis tool, provides the best means of representing risk, based on the above, and this is expressed as:

$$\text{Risk} = \text{criticality of the occurrence} \times \text{frequency of occurrence}$$

To this, detection can be used for risk mitigation. Various schemes enhance the risk assessment process by using a numerical scoring system so that one risk can be examined relative to another (which allows a risk to be considered as high, medium or low) [3]. Importantly, such risk schemas are highly generalized and, being applied cross industry, do not deal with the particularities of microbial contamination. The conception of risk as "the combination of the probability of occurrence of harm and the severity of that harm." Harm, in the pharmaceutical processing context, is damage to health, including the damage that can occur from loss of product quality or availability. With this latter point, the concern is not only with an adulterated medicine reaching the marketplace but also with an incident that prevents a pharmaceutical organization from releasing a medicine that consequently leads to a patients going without a much needed medicinal product. An important term in risk assessment is hazards. A hazard can be best defined as the potential source of harm.

Once risk is defined, an attempt can be made to understand if the risk acceptable and what controls are available to mitigate the risk. This can be understood by posing the following questions:

1. is the risk above an acceptable level?
2. what can be done to reduce or eliminate risks?
3. what is the appropriate balance among benefits, risks, and resources?
4. are new risks introduced as a result of the identified risks being controlled?

The pharmaceutical industry and regulators assess and manage risk using recognized risk management tools. A nonexhaustive list of some of these tools includes [4]:

1. basic risk management facilitation methods (flowcharts, check sheets, etc.);
2. FMEA;
3. failure mode, effects, and criticality analysis (FMECA);
4. fault tree analysis (FTA);
5. hazard analysis and critical control points (HACCP);
6. hazard operability analysis (HAZOP);
7. preliminary hazard analysis (PHA);
8. risk ranking and filtering;
9. supporting statistical tools.

These tools differ in structure and whether they are qualitative or quantitative in output. Nevertheless, they share commonality in approach in that they involve:

- identifying hazards;
- analyzing the risk associated with each hazard;
- evaluating how significant the risks are.

The correct identification of all of the hazards, their routes of transfer and their control methods and an accurate assessment of the degree of risk that these hazards provide is a fundamental stage in ensuring the effectiveness of any risk management system. Risk management is fundamentally about understanding what is most important for the control of product quality and then focusing resources on managing and controlling the important aspects in order to ensure that risks are reduced and contained. Before risks can be managed, or controlled, they need to be assessed, hence the centrality of "risk assessment."

A detailed evaluation of a risk management strategy is beyond the scope of this book; however, it is important to be aware of the initiatives within the pharmaceutical industry with respect to risk management and some of the tools available. The knowledge and understanding of the areas of the greatest risk associated with any manufacturing process enables resource and funding to be focused upon these areas, optimizing the security, and productivity of the process.

With risk assessment, whichever tool is adopted, there are two important points to bear in mind:

- there is no such thing as "zero risk," and therefore, a decision is required as to what is "acceptable risk." This must be qualified before the risk assessment begins;
- risk assessment is not an exact science—different people will have a different perspective on the same hazard.

The outcome of the risk assessment should be used to determine appropriate control strategies needed to reduce the risk to a level that is deemed acceptable. These strategies are typically focused on reducing the probability of a risk occurring. Where risk cannot be reduced down to a satisfactory level, then a mitigating strategy is to increasing the probability of the associated hazard being detected. An example is with environmental monitoring, where monitoring can be targeted to specific areas such as exposing settle plates near open product [5].

Risk assessments and the actions arising from them need to be documented and subject to periodic review to ensure that the assessment reflects the actions taken and the most recent data.

18.3   The need for microbiological risk assessment

Microbiological risk assessment should be carried out for both sterile and nonsterile manufacturing activities to establish what microbial risks are involved with the facility, equipment, and processes used. Whether such risk assessments should be activity specific (such as producing purified water, cleaning equipment, or freeze drying) or product specific, will depend upon the type of organization. In many cases, a generic risk assessment can cover a number of products; however, in doing so, product-specific considerations and the nature of operations should not be forgotten.

The outcome of a microbiological risk assessment can help determine the appropriate controls associated with the running of the facility or specific processes and an appropriate monitoring program.

When undertaking microbiological risk assessments, a description of the full manufacturing process is a useful starting point in assessing the risks involved. This may be simply done as a process map or flow diagram, with activities, equipment used and process parameters added on. Important factors to consider in the manufacturing process, and in relation to the nature of microorganisms, include:

- solvents used:
  - water-based processes provide a more favorable environment for microorganisms;
  - the use of other solvents might decrease the risk of microbial growth;
- pH:
  - values above 10 or below 2 generally being detrimental to microorganisms;
- osmolarity of solutions:
  - high osmolarity is typically detrimental to microbial growth;
- temperatures used:
  - a temperature range of 25–35 °C generally promotes microbial growth; significantly higher or lower temperatures are detrimental to most microorganisms. However, any risk assessment should consider common microbiota to see if psychrophiles or thermophiles (bacteria or fungi) are present;
- drying:
  - water is a vector and a growth source, areas should be kept as dry as possible and water supplies need to be controlled;
  - in relation to product contamination risk, if the water activity of the product is reduced below 0.6, then microbial growth will be suppressed [6]. A fundamental component of assessing the risk for microbiological control in nonsterile manufacturing is an understanding of whether the product or intermediates during the production process are able to support growth or sustain viability of microorganisms is water activity. Water activity

(aw) is a measure of the free water in a material and is, therefore, a useful measure to aid the determination of microbiological risk. Water activity is a more accurate index for microbial growth than water content as microorganisms have a limiting water activity below which they cannot grow (typically a level of 0.6, on a scale of 0–1.0). Water activity can vary according to different temperatures; thus, temperature control is important for assessing product risk in relation to storage;

- understanding water activity, in the context of risk, allows the microbiologist to: develop product formulations; set microbiological release specifications; establish microbial testing programs; and determine the potential shelf life stability from microbial growth;
- hold times and overall campaign length:
  - longer processing times may increase the opportunity for microbial proliferation unless the conditions are detrimental to microbial growth;
- open processing:
  - open processing is at greater risk compared with closed processing. Protective measures can be put in place through the use of unidirectional airflow and staff gowning and glove spraying;
  - in general, the longer the exposure period with open processing then the greater the likelihood of contamination transfer (as discussed below);
- fixed or mobile equipment:
  - equipment that has an in-build clean-in-place or sterilize-in-place capability is generally under less risk, provide that it has been qualified, than equipment that needs to be taken out of the process area and to a wash bay;
- personnel interaction [7]:
  - the higher the level of personnel activity then the greater the risk of contamination, given that people are the primary source of contamination in cleanrooms;
  - a related factor is the level of room occupancy;
- environmental control:
  - the level of environmental control, such as the way that the heating, ventilation and air conditioning system (HVAC) operates will affect how contamination is distributed or the rate at which it is removed from the critical area. Here an understanding of factors like air distribution, air change rates, pressure differentials, recovery times, filter efficiency, and so on is important;
  - environmental monitoring, although distinct from environmental control, can inform about the capability of environmental control systems;
- types of microorganisms:
  - the microbiologist must know and understand the microflora within the facility and trend the findings. Shifts in what is normally recovered can provide valuable information about risks. For example, a rise in the population of *Pseudomonad*-related genera may suggest a breakdown in water control. To take a second example, a rise in the population of *Bacillus* species could be linked to in-coming materials or the presence of dust; whereas Gram-positive cocci, such as a species from the genera *Micrococcus* or *Kocuria*, will probably have an association with personnel;
  - in addition, for terminal sterilization, the likely bioburden is not simply about microbial numbers for the types of species and theoretical resistance to the sterilization process will affect the success or otherwise of the sterilization cycle;
  - with aseptic processing, identifying microorganisms in different locations informs about the likelihood of contamination transfer and helps to identify points of origin (for example, linking a Grade A settle plate with a finger plate taken from an operator who performed a filling machine intervention);

- nature of microorganisms:
    - microorganisms in different states can survive for long periods. As the book has considered, microorganisms that can enter dormancy through forming endospores (such as *Bacillus* and *Clostridium* species) can survive in conditions that vegetative microorganisms cannot. The risk is elevated should endospores be given the opportunity to germinate;
    - furthermore, while the water activity, as indicated above, provides a useful indicator of the ability of the formulation to support growth, it should be remembered that some organisms present may remain viable and be pathogenic at low levels (such as *Salmonella* species);
    - with these examples it remains that good control during manufacturing is still essential.

For the manufacture or processing of nonsterile and sterile pharmaceutical products, such risk assessments can only be undertaken accurately if the correct models of contamination are understood and utilized.

This chapter proceeds to describe the fundamental mechanism of contamination transfer and details how this can be utilized to provide an effective assessment of the risk of contamination, from both microorganisms and microbial carrying particles, to pharmaceutical products.

18.4   Microbial contamination transfer

Important to microbial risk is the concept of contamination transfer. If microorganisms are deposited onto a surface some distance from a site of risk (such as exposed product), then the likelihood of product contamination is dependent upon the possibility of the microorganisms being transferred to the product. For this to happen, a vector is required. The common vectors are air, water, or physical movement (such as via the hands of a production operative) [8].

The chance of microorganisms being transferred from a source of contamination to a product is dependent on the likelihood of them being dispersed from the source, transferred to and then deposited onto or into a product. Where contamination transfer occurs, the important variable is time, and the microbiologist needs to account for the number of microorganisms deposited onto a given area in a given time. An attempt to express this phenomenon as a universal equation has been made by Whyte and Eaton [9]:

$$\text{Contamination deposited on a product} = C \times S \times P_d \times P_a \times A \times T$$

where C is the concentration of contamination on, or in, a source (number/cm^2 for a surface, or number/cm^3 for air); S is the quantity of surface material, or air, that is dispersed or transferred from a source in a given time (cm^2/s for surfaces, and cm^3/s for air dispersion); P_d is the proportion of contamination dispersed from a source that are transferred to the area adjacent to the product; P_a is the proportion of contamination in the adjacent area that are deposited per unit area of the product (/cm^2); A is the area of surface onto which the contamination is deposited (cm^2); and T is the time, during which transfers occur (seconds).

The equation expresses the fact that the amount of microbial contamination is dependent on:

1. the concentration on a contaminating surface, or within air;
2. how much of this contamination is dispersed, transferred, and deposited onto, or into, the product;
3. a variable of time or event frequency.

A practical example of the mechanisms outlined would be the airborne transfer of skin microorganisms from personnel to a product. Here, the number of microorganisms that would deposit on a product is dependent upon:

1. the concentration of microorganisms per area of a skin surface;
2. the surface area of skin that is dispersed in a given time;
3. the proportion of microorganisms dispersed and transferred through cleanroom clothing and the cleanroom air to the area adjacent to the product;
4. the proportion of microorganisms adjacent to product that will be deposited onto a given area of exposed product;
5. the time over which this deposition occurs.

While this general model is useful for conceptualizing risk, for airborne and surface contamination, assessing microbial risk sometimes requires more detailed discussion. These areas are considered next.

18.4.1 Airborne deposition

Most of the microorganisms found in cleanroom air derive from the skin of personnel within these areas. A proportion of these skin cells carry microorganisms, and cleanroom personnel can disperse several hundred microbial carrying particles per minute through cleanroom clothing. For these reasons, microorganisms are normally found in cleanrooms attached to skin particles (or very occasionally, a clothing fiber). The average size of microbe carrying particles will vary between about 8 and 20 μm and can deposit, mainly by gravity into, or onto, the product [10].

The number of microbial carrying particles that will deposit onto a given area of product, and hence the proportion of product contaminated can be calculated using the deposition rate obtained from microbiological settle plates. This approach, also first postulated by Whyte, can be used for most pharmaceutical-manufacturing process where airborne contamination deposits passively from the air, mainly through the force of gravity, into, or onto, the product [11].

The general equation (above) can be modified by combining the first four variables of the equation into a single variable that provides the number of microorganisms that will deposit onto a given area of product in a given time (termed the deposition rate). This can be expressed as:

$$\begin{array}{l} \text{Number of airborne microorganisms} \\ \text{deposited onto a product in} \\ \text{a given time (deposition rate)} \end{array} = \begin{array}{l} \text{area of product} \\ \text{exposed} (\text{cm}^2) \end{array} \times \text{time of exposure}(\text{s})$$

With the answer expressed as number of microorganisms per square centimeter at a given rate of time.

Therefore, in cleanrooms settle plates can provide information about the deposition rate and thus the likelihood of product contamination over a period of time. In addition, by capturing microorganisms, settle plates allows the contaminated to be characterized, which helps with assessing the point of origin.

18.4.2 Surface contact

The equation used to calculate the number of microorganisms deposited by surface contact is also derived from the general equation. This risk is calculated by considering surface contamination as something that occurs as discrete events, and by combining the dispersion, transfer, and deposition variables into one overall term, that is, the "transfer coefficient." This equation can be reformatted to the following equation that calculates the number of microbes deposited on a given area of product over a known time. The equation is:

$$
\begin{array}{l}
\text{Number of} \\
\text{microorganisms} \\
\text{deposited by} \\
\text{surface contact} \, (\text{no.})
\end{array}
=
\begin{array}{l}
\text{microbes or} \\
\text{particles on} \\
\text{source surface} \\
(\text{no.} / \text{cm}^2)
\end{array}
\times
\begin{array}{l}
\text{transfer} \\
\text{coefficient}
\end{array}
\times
\begin{array}{l}
\text{area of} \\
\text{contact} \, (\text{cm}^2)
\end{array}
\times
\begin{array}{l}
\text{frequency of} \\
\text{contact} \, (\text{no.})
\end{array}
$$

The "transfer coefficient" represents the proportion of microorganisms on the source surface that are transferred onto the product with each contact. A practical example would be contacts of a finger with the surface of a product. The number of microbes transferred could be calculated from (a) the concentration of microorganisms on the finger surface, (b) the proportion of microorganisms on the finger surface that are transferred to the product (i.e., the transfer coefficient), (c) the area of contact between the finger and product, and (d) frequency, that is the number of times the product is touched.

18.5 Identification of sources and routes of contamination

The key step in the risk assessment process is the identification of the sources of microorganisms, their routes of transfer and control methods. All sources of contamination must be identified and their risk to the product assessed. With this, grouping of the sources and the use of a risk diagram are useful.

18.5.1 Sources of contamination

Examples of sources of contamination in a typical cleanroom are as follows:

- adjacent areas;
- supply air;
- cleanroom air;
- surfaces;

- people;
- machines;
- ancillary equipment;
- materials;
- containers;
- packaging;
- liquids.

Areas adjacent to product processing areas, such as change rooms, transfer hatches, and external corridors, are likely to be more contaminated than the cleanroom or controlled area used for processing. The transfer hatches and change areas will be contaminated by the activities in these areas, and there may be less control of the dispersion of contamination in the outside corridors. The transfer of this contamination into the production cleanroom should be minimized. The air supplied to a cleanroom, if not correctly filtered, is a source of contamination and the air within the cleanroom is a major source of contamination and contains contaminants dispersed from people and machinery.

The floor, walls, ceilings, and other surfaces in the cleanroom, such as tables, tools, paper, and so on, are fair examples of sources of surface contamination which is normally derived in a secondary way from personnel touching them, or from contamination deposited from the air. These surfaces can also be primary sources of contamination if poor quality constructional components have been used, as they break up and disperse fibers of wood or particles of plaster.

As discussed, personnel within the cleanroom are a major source of contamination. People can disperse vast quantities of contamination from the skin, hair, and mouth. This contamination can be transferred to the product through the air, or by contact with their hands or clothing. Cleanroom clothing, gloves, and masks are used to control the contamination being dispersed from the people wearing them. These items of clothing can, however, become contaminated by the people wearing them and from other cleanroom sources [12].

Machines are a source of particles, as they can generate contamination by the movement of their constituent parts, or a secondary source from contamination deposited on them from the air or by contact with personnel. Ancillary equipment such as chairs, air samplers, and calculators will have contamination similar to the surfaces previously considered. Raw materials, containers, and packaging that are transferred into the cleanroom may be contaminated and liquids, such as those used for the product formulation, may also be a source of contamination.

18.6 Routes of transfer

The routes of transfer of contamination, by airborne and surface contact, should also be identified because by minimizing these routes, the risk of contamination can also be reduced.

Airborne contamination is normally sourced from people and machines and is dispersed into the air and then deposited onto the product. If the particles are small, like skin cells, they can move around in the air before depositing. However, if they are large particles, like spittle, dandruff, or cuttings of plastic or glass, they will remain within a short distance from where they were generated, and fall directly into, or onto, the product; this is called intimate airborne spread.

Contact contamination occurs when contaminated items such as machinery, ancillary equipment, cleanroom surfaces, containers, packaging, gloves, and clothes come into contact with the product. Contact contamination can occur in many ways; one example is when personnel touch a contaminated surface with their gloves, which then become contaminated. If product is then touched with that glove, contamination is transferred onto the product.

Using information of the type discussed above, the sources and routes of transfer of contamination can be determined, especially when superimposed over a process risk diagram.

18.7 Risk assessments for general cleanroom areas

Risk to cleanroom products from surface contact and airborne deposition contamination can be assessed, either at the preliminary design stage of the cleanroom and associated manufacturing process or, retrospectively, for an established manufacturing operation.

All microbial sources, such as those outlined earlier, should be considered to be potential hazards and assessed to determine their degree of risk. The likelihood of deposition of contamination onto or into a product is very much dependent upon factors associated with the product itself, such as the exposed area and the time of exposure. In order to compare the hazards in the cleanroom areas on an equal footing, the variable of deposition of microorganisms is assumed to be constant and can, therefore, be ignored for this particular assessment.

Furthermore, the variable of frequency maybe continuous, as in the case of the hazard associated with air supplied to the cleanroom areas, or it may be associated with the transient transfer of contamination by personnel during the manufacturing operation. The variable of frequency is, therefore, also not utilized. Therefore, a version of the fundamental equation should be used for this assessment. This is:

$$\text{Risk from microbial contamination}\left(\text{risk rating}\right) = A \times B \times C \times D$$

where A is the microbial contamination on, or in, a source; B is the ease of dispersion of contamination from the source; C is the ease of movement of contamination to product; and D is the proximity of contaminating source from the product.

The risk rating of each source of contamination can be determined by assigning risk scores to the risk factors A–D. It should be noted that these risk factors, and the associated risk scores, take into account the measures that have been utilized to control the identified hazards and therefore the resultant risk rating relates to the risk in the controlled (operational) state. This method can also be used to assess the level of risk in the uncontrolled or partially controlled state simply by re-assessment of the risk score and re-calculation of the risk rating for the level of control employed. This approach of considering the risk in the controlled state has found to provide the best and most flexible method for this type of overall risk assessment.

18.8 Risk scoring systems

Risk scores are usually assigned to hazards and an associated scoring method must be established. It is easier to describe risk by simple words modified to denote greater or lesser importance and to then allocate a score to these words. The most accurate scoring system must also have the meaning of the word descriptions to be in direct proportion to the score magnitude, and it should span the whole range of the risks considered. Three possible systems are shown in Table 18.1.

An example of how risk scores can be allocated to different risk factors is shown in Table 18.2. This example uses the five-stage scoring system.

With Table 18.2, for each identified contamination source, the risk scores for each risk factor should be determined and then multiplied together (the fundamental equation shows that multiplication and not addition is required) to obtain a risk rating. This risk rating determines the degree of risk associated with each contamination source.

Keeping with this approach, general sources of contamination and calculated risk ratings are shown in Table 18.3.

Alternatively, in place of a numerical system, the risk rating can be assigned a "low," "medium," or "higher" category.

Table 18.1   Risk scoring system

7 Stage		5 Stage		4 Stage	
Not possible	0	Nil	0	Nil	0
Very unlikely	1	Very low	0.5	Low	1
Unlikely	2	Low	1	Medium	2
Possible	3	Medium	1.5	High	3
Likely	4	High	2		
Very likely	5				
Definite	6				

Table 18.2   Scores for risk factors used for assessing hazards

Risk factor (A) Amount of microbial contamination on, or in, a source	Risk factor (B) Ease of dispersion of microorganisms from the source	Risk factor (C) Ease of movement of contamination to product	Risk factor (D) Proximity of source from product
0 = nil	0 = nil	0 = nil	0 = remote
0.5 = very low	0.5 = very low	0.5 = very low	0.5 = outside corridor, air lock
1 = low	1 = low	1 = low	1 = periphery of cleanroom
1.5 = medium	1.5 = medium	1.5 = medium	1.5 = general area of cleanroom
2 = high	2 = high	2 = high	2 = critical area

Table 18.3 A general model for assessing risk within pharmaceutical processing areas

Source	Microbial counts Surface- counts/24cm²; air-counts/m³	Risk factor A Conc. of microbes	Risk factor B Ease of dispersion from source	Risk factor C Ease of movement to product	Risk factor D Proximity to product	Risk rating
1. Areas adjacent to production cleanrooms (change and transfer areas)						
1.1 Air outside production cleanrooms	>10 (terminal air filtration, air supply rates)	1.5	2	0.5 (differential air pressure, physical barriers)	0.5	0.75
1.2 Floor surfaces outside production cleanrooms	>5 (floor disinfection, overshoes, tacky mats)	1.5	1	0.5 (physical barrier, footwear change)	0.5	0.38
2. Supply air						
2.1 Air supplied to unidirectional air flow areas	<1 (terminal air filtration)	0.5	2	2 (unidirectional air flow)	2	4
2.2 Air supplied to turbulently ventilated cleanroom	<1 (terminal air filtration)	0.5	2	0.5 (physical barrier, air dilution with turbulent airflow)	1.5	0.75
3. Air within unidirectional air flow areas and cleanrooms						
3.1 Unidirectional air flow areas	1 (garments, unidirectional airflow velocity)	1	2	0.5 (unidirectional airflow)	2	2
3.2 Turbulently ventilated cleanroom	10 (garments, air supply rates)	1.5	2	0.5 (physical partition, unidirectional air flow)	1.5	2.25
4. Machines and ancillaries						
4.1 Machine surfaces not in contact with product	1 (surface disinfection, environmental control)	1	0.5	1 (aseptic behaviors)	2	1
4.2 Machine surfaces in direct contact with product	<1 (sterilization)	0.5	0.5	2	2	1

4.3 Ancillaries (product scissors, forceps, etc.) in direct contact with product	<1 (sterilization)	0.5	2	2	2	1
5. Non-machine surfaces						
5.1 Ceilings, walls, floors, doors in turbulently ventilated cleanroom	>1 (disinfection, aseptic behaviors, garments)	1 (aseptic behaviors)	0.5	1 (aseptic behaviors)	1.5	0.75
5.2 Trolleys, chairs, tables, eyewash, calculator, waste bins, paperwork, pens, bin bags, labels, press buttons, and switches etc. in cleanroom	1 (surface disinfection, glove disinfection)	1	0.5	1.5 (aseptic behaviors, glove disinfection)	1.5	1.1
5.3 Walls, floors, and ancillaries; and for microbial samplers, located in unidirectional air flow area	1 (surface disinfection, aseptic behaviors, garments)	1	0.5	2 (aseptic behaviors, glove disinfection)	2	2
6. People						
6.1 Transfer to product via gloved hands [1]	<100 (hand washing)	2	2	0.5 (2 pairs of gloves)	2	4
6.2 Transfer to product via gloves with secondary contamination [1]	<1 (glove disinfection, aseptic behaviors)	0.5	2	2 (aseptic behaviors)	2	4
6.3 Airborne transfer of microorganisms from personnel working in unidirectional air flow area	>2000	2	2	0.5 (garments, aseptic behaviors, unidirectional air flow)	2	4
6.4 Surface transfer to product from cleanroom clothing	>1 (garments and aseptic behaviors)	1	2	1 (aseptic behaviors)	2	4
7. Material - primary and packaging						
7.1 Liquid product from clean process area	<1 (sterile filtration)	0.5	2	2	2	4
7.2 Container	<1 (sterilization)	0.5	2	2	2	4

The magnitude of the risk ratings can be used to determine the degree of effort to be allocated into controlling and monitoring each source. However, it should be appreciated that the risk assessment method should only be used to *assist* in assessing the risks. The quality of the input information and the inexact nature of the model ensure that exact predictions cannot be made.

18.9   Conclusion

This chapter has presented an introduction to the important subject of risk in biopharmaceuticals and with microbiological risks in particular. Understanding where microorganisms may reside, together with the typical types of organisms helps with building in risk mitigation into pharmaceutical and healthcare processes. What is also of importance is mapping out the possibility of contamination transfer through air-streams or direct transfer through personnel. Knowing the likelihood and severity of such risks is important for an assessment of environmental monitoring. As well as these proactive measures, risk assessment is also helpful for dealing with contamination events. While risk assessment is a complex and wide-ranging subject, this objective of this chapter was provider a lead in to the subject and to assist those who need to understand the fundamentals of contamination control.

References

[1]   Sandle T, Lamba SS. Effectively incorporating quality risk management into quality systems. In: Saghee MR, editor. Achieving quality and compliance excellence in pharmaceuticals: a master class GMP guide. New Delhi: Business Horizons; 2012. p. 89–128.
[2]   Sandle T. Risk management in pharmaceutical microbiology. In: Saghee MR, Sandle T, Tidswell EC, editors. Microbiology and sterility assurance in pharmaceuticals and medical devices. New Delhi: Business Horizons; 2011. p. 553–88.
[3]   Sandle T. Environmental monitoring risk assessment. J GXP Compliance 2006;10(2):54–73.
[4]   Sandle T. The use of a risk assessment in the pharmaceutical industry—the application of FMEA to a sterility testing isolator: a case study. Eur J Parenter Pharm Sci 2003;8(2):43–9.
[5]   Whyte W. In support of settle plates. J Pharm Sci Technol 1996;50:201–4.
[6]   Kabara JJ, Orth DS. Preservative-free and self-preserving cosmetics and drugs: principles and practise. New York: Marcel Dekker; 1997, p. 1–14.
[7]   Whyte W, Bailey PV. Reduction of microbial dispersion by clothing. J Parenter Sci Technol 1985;39:51–60.
[8]   Sandle T. Contamination control risk assessment. In: Masden RE, Moldenhauer J, editors. Contamination control in healthcare product manufacturing, vol. 1. River Grove, IL: DHI Publishing; 2013. p. 423–74.
[9]   Whyte W, Eaton T. Microbiological contamination models for use in risk assessment during pharmaceutical production. Eur J Parenter Pharm Sci 2004;9(1):11–5.
[10] Noble WC, Lidwell OM, Kingston D. The size distribution of airborne particles carrying microorganisms. J Hygiene 1963;61:385.
[11] Whyte W. Sterility assurance and models for assessing airborne bacterial contamination. J Parenter Sci Technol 1986;40:188–97.
[12] Eaton T. A safe pair of hands—how secure are your gloves used for aseptically prepared pharmaceutical products? Eur J Parenter Pharm Sci 2005;10(3):35–42.

Manufacturing and validation

19

19.1 Introduction

In all pharmaceutical microbiology control laboratories, the frequency of "failure" in product related testing is exceedingly low. This is so with finished product testing, intermediate testing, and, in sterile manufacture, for environmental monitoring. With water testing, periodic out-of-limits will occur, as there will be with some starting materials; and with environmental monitoring of lower grade cleanrooms and within nonsterile facilities, there will be occasional excursions, especially from surface contact plates. However, overall microbial data deviations represent a small proportion of the collected total number of samples.

This state of control exists because manufacture is performed in equipment and facilities which have been hygienically designed, and which are operated and maintained according to hygienic principles. This is the consequence of successful application of the principles of good manufacturing practice (GMP) and quality assurance.

The basic principle of both good manufacturing practice and quality assurance is that only by having properly designed and operated processes can it be possible to obtain satisfactory product from unit to unit within a batch, and from one batch to other batches manufactured using different equipment and/or on different occasions. Achieving this requires that manufacturing facilities, equipment, and processes should be validated prior to being released for routine use; and subsequently in routine use they should always be operated to procedures accurately reflecting the conditions shown to be effective in validation.

This chapter focuses on two broad topics: manufacturing procedures and validation. It is important that the pharmaceutical microbiologist understands how manufacturing procedures and validation under-pin all aspects of quality and are, therefore, not wholly divorced from a requirement for microbiological input. In risk assessment terms, this reduces the severity of a hazard and the probability of that hazard occurring, with microbiological testing functioning as the detection tool (risk assessment terms are defined in Chapter 18).

19.2 Manufacturing procedures

GMP provides an important structure from which manufacturing procedures are shaped. Some key GMP elements are [1]:

- specifications describe in detail the requirements with which the products or materials used or obtained during manufacture have to conform;
- manufacturing formulae, processing and packaging instructions state all the starting and packaging materials used and, additionally, lay down all processing and packaging operations;

Pharmaceutical Microbiology. http://dx.doi.org/10.1016/B978-0-08-100022-9.00019-0

- procedures give directions for performing certain operations, for example, cleaning, clothing, environmental control, sampling, testing, and equipment operation;
- records provide a history of each batch of product including its distribution, and also of all other relevant circumstances pertinent to the quality of the final product.

These documented aspects will now be examined in more detail.

19.2.1   Specifications

Specifications may be organized and laid out differently from company to company, but essentially they all must contain the same elements. These are [2]:

(a) The identity of what is being specified. This should be unambiguous, but also intelligible. A name is usually accompanied by a code number or a part number. The identity of starting materials may be according to a pharmacopoeia or they may be simple or complex chemical molecules. Specifications for most chemicals should be accompanied by a reference to a method by which their identity should be confirmed. The identity of a finished product should refer to the concentration of the active ingredient(s) and other information pertaining to its registered formula; specifications for finished products are in this respect usually more detailed than specifications for starting materials;

(b) Limits on impurities or defects. These may be chemical or physical. Microbiological contamination is strictly speaking an impurity;

(c) Other characteristics determined to be of importance. It may be that the particle size of a starting material is of importance as to how it runs on a piece of equipment, or how it binds to form a tablet, or how it forms an emulsion in a cream or ointment, or even how the finished drug product performs therapeutically as in inhalation products. It may be that the pH of a finished injection product affects its therapeutic effects.

Specifically for pharmaceutical microbiology, limits on impurities or defects are of the greatest importance. These can be specified in a variety of ways. In the context of starting materials, these limits should be specified to protect the quality specifications of the finished product into which the materials are being incorporated. Attention is importance here since the cost of rejecting a batch of starting materials is considerably less than the cost of rejecting a batch of finished product.

For instance, with the case of an oral liquid in aqueous solution comprising some colorants, flavorings and preservatives together making up less than 1%, plus an active at 2%, and having a finished product specification of not more than 10^3 microorganisms per milliliter. The major starting material is purified water. The microbiological limit of not more than 10^2 microorganisms per milliliter placed on purified water is quite adequate to protect the finished product's microbiological specification even if all of the other starting materials were to be specified at the normal level for starting materials of not more than 10^3 microorganisms per gram or milliliter.

In contrast, considering a syrup, again with an active at 2%, but containing 80% sucrose. If the sucrose were to have a specification of not more than 10^3 microorganisms per gram there would be very little protection afforded to the finished product's microbiological specification. In this case, the specification for the sucrose would have to be sensibly tightened.

Defects in packaging materials may be specified with associated acceptable quality levels (AQLs). These are expressions of the worst quality level that is still considered satisfactory. Different AQLs may be applied to different defects according to their criticality. For instance, the inside diameter of the neck of a glass vial is likely to be a critical quality characteristic because of its potential effect on the maintenance of sterility, but glass flaws may be defined only as cosmetic quality characteristics and thus have "weaker" AQLs. The AQL is a statistical concept which when used in association with published tables defines the number of items that should be sampled and tested and how many defective items may be tolerable within a particular sample size.

It is customary that each batch of finished pharmaceutical products is sampled and tested against its specification for purposes of deciding it is suitable for release. For microbiological specifications, this is not always the case:

(a) where parametric release has been allowed for terminally sterilized pharmaceutical products, the test for sterility may be omitted;
(b) it is very unusual (even though limits may be registered) for batch by batch microbiological release testing to be applied to solid oral dosage forms such as tablets unless they contain high proportions of starting materials of plant or animal origin or known to carry high levels of contamination. This elimination or reduction of testing for specific dosage forms is a matter of professional judgement and risk and should always be justified and documented in the company's procedures;
(c) similarly, it is very unusual for batch by batch microbiological release testing to be applied to nonsterile products containing antibiotics. Once again, the justification for allowing this should always be justified and documented in the company's procedures.

For starting materials, GMP typically requires that there should be appropriate procedures or measures to assure the identity of the contents of each container of starting material. This is typically via an identity test. This applies to material identity but not necessarily to impurities, defects, or other quality characteristics. Testing of these characteristics is a matter of judgement and risk, and justifications should be documented [3].

19.2.2 Batch manufacturing records

Global GMPs are very specific about the contents of batch manufacturing records (BMRs). A BMR is a description of the milestones along the critical path of manufacture against which manufacturing personnel identify who did what, when they did it, what they did it with, and the critical measurements they made.

Traceability is the key element of the BMR. BMRs are controlled documents, one blank copy should be issued for each batch scheduled for manufacture (or released if a computerized system is used) [4]. This document should be completed and returned to the quality department for checking. The documents should be up-to-date and reflect what actually goes on with processes [5].

Pharmaceutical microbiologists are involved with some parts of the BMR. These include:

(a) Sterilization records are generally included in BMRs and in some companies may be diverted to the pharmaceutical microbiologist for checking. If sterilization records are not checked by the pharmaceutical microbiologist, it may well be the microbiologist's responsibility to ensure that whoever is performing this essential function has proper training in sterilization science and technology;

(b) In aseptic manufacture of sterile products, there may be a BMR for media fills. Within this BMR, the microbiologist may be required to complete the sections pertaining to verification of the growth support properties of the media, and to incubation and inspection of the media-filled containers.

Outside of these examples, it is generally only when there are problems, either in environmental microbiology or in failed finished product testing that the pharmaceutical microbiologist would be expected to encounter the BMR as a part of an investigation into potential manufacturing causes of the problem.

19.2.3   *Manufacturing standard operating procedures (SOPs)*

Whereas BMRs describe the milestones along the critical path, they do not provide the detail required to define adequately how the various tasks making up the manufacturing process should properly be done. For example, consider set up of an aseptic filling machine which may in the BMR detail only:

* names of the personnel undertaking the set up;
* start time;
* finish time;
* sterilization records of the parts installed.

When it is considered that set up may involve two or three personnel and the difference between the start time and the finish time is unlikely to be less than 30 min and possibly as long as 90 min, it can be appreciated that this is in fact a very complex procedure indeed.

Such complex procedures must be defined and decided and agreed by management through standard operating procedures. The pharmaceutical microbiologist should have an involvement in all manufacturing SOPs that impact on hygiene. This could include, for instance, water system operation, clean-in-place cycles, and process water systems.

19.3   Validation

There are varying definitions of validation within the pharmaceutical setting. One of the clearest is provided by Agalloco, who writes [6]:

> *Validation is a defined program, which in combination with routine production methods and quality control techniques, provides documented assurance that a system is performing as intended and/or that a product conforms to its predetermined specifications. When practiced in a lifecycle model, it incorporates design, development, evaluation, operation and maintenance considerations to provide both operating benefits and regulatory compliance.*

Validation is a critical concept in the pharmaceutical manufacturing industry. There are some key quality attributes, which include [7]:

(a) Documented evidence: validation is an activity which must be recorded and be formally documented for inspection;

(b) Consistently: a process that cannot be shown to be capable of performing consistently in the manner intended is of little value to manufacturing. One of the tenets of validation is that before a process is released for routine manufacture, it should have been shown through a sufficient number of replicate trials that it is capable of performing consistently;

(c) Pre-determined: the expression predetermined when applied to specifications and quality attributes indicates that validation is a confirmatory exercise and not an exploratory one. The exploration involved in new manufacturing processes belongs with the concept of development; it can be determination of what works (and preferably why it works), what its bounding limits (parameters) may be, and even process optimization, but it is not validation. Validation follows only when the limits have been predetermined. Hence, validation is confirmatory.

Good validation is well planned. Thus, the validation program should be defined and documented in a validation master plan (VMP) or equivalent documents. VMPs commonly contain [8]:

(a) *Application and scope.* The VMP must describe unambiguously what is being validated (the application) and the scope of the validation. For instance, it might be that the application is validation of blending for a particular tablet product, using a specific blender located in a particular blending room (the scope). It may be that the blender has been in previous use in that location and some qualifications have been already done in connection with other validations. In such a case some, but possibly not all, qualifications may not need to be repeated but may merely be referenced. On the other hand, it may be that the qualifications done previously need to be re-examined and perhaps updated. The definition of these activities belongs in the VMP;

(b) *Qualifications.* In some instances, it may be practical to combine qualifications, and if this is the case, it should be stated in the VMP. It is universally the case that all new processes are prospectively validated (i.e., all qualifications must be complete before the process is released for routine use), but there may be items of equipment or services (e.g., steam generators, compressors, etc.) identified when preparing the VMP that have not been previously qualified. If this is the case, their retrospective qualification should be included in the VMP;

(c) *Standards.* The VMP is not the place for detailing the standards and limits being applied, this would merely amount to a repetition of detail necessarily included in the qualification protocols. However, it might be that the standards being applied differ, and the VMP is the place to state which standards apply;

(d) *Deviations.* Any deviations that occur need to be addressed and signed off before the validation itself is allowed to be completed;

(e) *Disposition of materials.* The VMP should define if product (or intermediates) manufactured as a part of validation trials may be released to market, and under what conditions this might be possible;

(f) *Revalidation.* Validation does not stop with the final sign off on the VMP. The philosophy of validation is that it continues through the "lifecycle" of an item of equipment or a process.

Generally, validation activities are structured in the same way. Here, there are a series of qualifications that must take place sequentially; each qualification must be completed and signed off before the subsequent one is allowed to begin, and all must be completed and in place before the final validation can be approved and the process released for routine use. Each qualification comprises a protocol predetermined and approved before the work is allowed to begin, reflected by a report on which the actual results obtained are recorded.

Validity should be reviewed periodically (validation review) through scrutiny of equipment logs, maintenance records, deviations, out-of-specifications, and periodic product quality review reports to determine if the equipment or process is still operating consistently to the same predetermined specifications and quality attributes. If not there may be some further requirement for process development or equipment modification or even withdrawal from use. The frequency for formal validation review should be defined in the VMP [9].

Mostly "re-validation" in the sense of repeating some aspect of the original validation (usually performance or process qualification) is only required in the event of a significant change, or from something highlighted in validation review [10]. However, there are some processes and items of equipment (notably sterilization processes, autoclaves, ovens, tunnels, etc.) that require a regular periodic re-qualification. Mostly, the processes requiring re-validation have been identified by the regulatory authorities either in guidance documents or through custom and practice at inspection. They have been determined from risk analysis (probably intuitively rather than by formal risk analysis) and are largely in those areas where serious patient risk could arise from undetected or undetectable "slippage" in the performance of a piece of equipment or a process.

19.3.1   Qualifications

As a part of the validation approach, there are a series of qualifications that form a part of the process [11]. These are discussed below.

(a) User requirement specification (URS). For any new project, there has to be a URS. Some URSs may never lead on to validation. This is because the URS is an expression of what a potential user of a new piece of equipment wants, but the item of equipment may never be approved and purchased. The URS predates the VMP.

For instance, when the pharmaceutical microbiologist needs a new laboratory autoclave, they should define the loads that they want to sterilize; the size of the device; and the means of its operation.

Once a URS exists and an approval in principle to purchase is obtained, functional specifications or designs may be obtained from various suppliers for whatever has been identified. Once it is decided what is going to be purchased, a VMP can be launched, and the validation program formally begins [12].

(b) Design qualification (DQ). DQ compares the functional specification or the design to the URS. A DQ protocol can be as little as a "tick list" reflecting the content of the URS.

If the URS has been prepared properly, DQ will reflect the compromises that are necessary in the "real world." Nothing "fits perfectly," but it might fit well enough. The purpose of DQ is to determine where the compromises may have to be made and whether they are acceptable.

(c) Installation qualification (IQ). IQ is the process of verifying that what the user believed they were buying is really what you got. IQ protocols identify the key elements of the specification or design. For instance, if an item of equipment was specified to be made from 316L stainless steel a metallurgy certificate needs to be provided alongside the piece of equipment.

(d) Factory acceptance test (FAT). If the company is purchasing a major piece of equipment, say an autoclave from Italy, it makes no sense to wait until it arrives in the warehouse before verifying that it has been built to the correct specification. For this, FAT is permitted; but this testing has to be done under the supervision of the purchaser's representatives, using the purchaser's documentation. An alternative option is site acceptance testing (SAT) in which verification is done in an engineering workshop or a warehouse rather than on the manufacturing floor.

(e) Operational qualification (OQ). OQ addresses whether the piece of equipment is capable of performing in the manner intended over the operating range intended, when installed and supported by local services. It also embraces calibration of measuring devices, establishment of maintenance procedures and schedules, training of maintenance operators, establishment of operating procedures, and training of production operators. The OQ may in some cases be conveniently amalgamated with IQ.

It is commonplace that, apart from the qualification of a new piece of equipment, there may also be a need to investigate how it operates best and to "optimize" it. Although this work may be being done on the same equipment and in the same broad timeframe as OQ it should be regarded as a development exercise and recorded as such.

(f) Performance (process) qualification (PQ). The final qualification, PQ, involves production materials or validation batches. It is PQ that calls for replication (as stated above, usually three times up to now, but who knows how many times in the future except that the number should be a function of risk and is unlikely to be less than three).

The process conditions must be defined and complied with during PQ, and the output must comply with its specifications. If PQ fails to comply with its acceptance criteria, either with respect to the process parameters or with respect to the product specification, or both, there is no choice but to re-develop the process through modifications either to the equipment or to the ways of operating the equipment.

19.3.2 Cleaning validation

A specialized area of validation, and of importance to pharmaceutical microbiology and to contamination control, is cleaning validation. This topic acts as a concrete example of how validation and microbiology interact. Cleaning validation is about providing proof of the effectiveness of the ways in which items of manufacturing equipment are cleaned. This presupposes of course that a cleaning process has been defined.

There are various levels of risk associated with cleaning validation [13]:

• Cleaning between different products. The consequence of carrying an active pharmaceutical ingredient into a second product which should not contain it is that the second product becomes adulterated. This is obviously serious and is a major risk in multiproduct equipment;
• Cleaning between batches of the same product. The risks here are lesser than between different products, but impurities and break down products can be carried over;
• Carry-over of cleaning agents. A self-created problem, but a problem nonetheless; indeed one which could be more consequential than carry-over between batches of the same product;
• Presence and survival of microorganisms.

The effort required to be included in cleaning VMPs and protocols is a reflection of the risk level involved.

Cleaning validation customarily requires selection of the "most difficult to clean product" among a range used on multiproduct equipment. Removal by cleaning should be tested for by product-specific methods; limits on residues can be calculated from advice contained in the various regulatory guides on the topic. Suspensions and emulsions are generally regarded as being among the most difficult to clean products. High solubility in water often makes cleaning easier.

The second important decision to make and document in cleaning VMPs is where to sample for residues—the "most difficult to clean locations." Dead-legs in pipework, areas beneath valve seatings, and so on are amongst typical locations. Direct swab samples are preferred to rinse samples.

Microbiological considerations are required by the regulations to be included in cleaning validation. Arguably, microbiological considerations should consist largely of preventative measure rather than removal of contamination once it has occurred. This points directly to an expectation that the pharmaceutical microbiologist's role extends beyond the laboratory, really to the URS.

Moreover, routine cleaning and storage should not allow microbiological proliferation. Drying after cleaning is perceived to be the most important aspect of preventing proliferation. A good approach emphasizes that even if cleaning is followed at some time later by a sterilization process, there is the attendant risk that microorganisms may have proliferated to the extent that unacceptably high levels of endotoxin/pyrogens may remain on the equipment, and by surviving the sterilization process come to contaminate products manufactured on the equipment.

19.3.3 Other validation exercises requiring the involvement of pharmaceutical microbiology

Potentially, pharmaceutical microbiology could be involved in all validations. This section briefly validates where pharmaceutical validation is important.

(a) Design of new facilities (and modification of existing facilities). Hygiene is a critical quality of all pharmaceutical manufacturing facilities, not just those dedicated to manufacture of sterile products. Pharmaceutical microbiologists should contribute to HVAC design, facility design and to the selection of materials versus disinfectant activity, etc.;

(b) Installation of new water systems (and modification of existing water systems). There are very few problems with water systems meeting the chemical and physical properties of pharmacopoeial grade waters. Microbiological problems are almost inevitable except where high temperature storage and distribution systems have been installed;

(c) Installation of new thermal sterilization processes;

(d) Introduction of new processes involving bacteria-retentive filtration. The microbiological qualification of bacteria-retentive filters is a highly specialized job, generally undertaken by the filter suppliers. Nonetheless, the pharmaceutical microbiologist should be involved in verifying that this has been done and may well be the person who has to explain it at inspection;

(e) Process qualification of all products with a significant water content. Microorganisms need water to grow and increase in numbers. Conversely, wherever there is water, there is also the potential for microorganisms to grow and increase in numbers. The types of microorganisms that grow in aqueous environments generally have extraordinary biochemical properties that allow them to metabolize complex pharmacologically active molecules, have the capability of causing infections in even healthy patients, and have the ability when present as an infection, to resist antibiotic therapy.

19.4   Conclusion

This chapter has examined two important aspects of the modern pharmaceutical plant: documentation and structure of manufacturing, and the steps required with process validation. With each of these, there is a very important role for the pharmaceutical microbiologist, not least in ensuring that adequate steps are being taken in relation to contamination (and that these have been reliably demonstrated). The foremost way to achieve this is through risk assessment, applying the principles and approaches discussed in Chapter 18.

References

[1]   Abdellah A, Noordin M, Wan Ismail WA. Importance and globalization status of good manufacturing practice (GMP) requirements for pharmaceutical excipients. Saudi Pharm J 2015;23(1):9–13.

[2]   Muselík J, Franc A, Doležel P, Gonĕc R, Krondlová A, Lukášová I. Influence of process parameters on content uniformity of a low dose active pharmaceutical ingredient in a tablet formulation according to GMP. Acta Pharm 2014;64(3):355–67.

[3]   Sandle T, Lamba SS. Effectively incorporating quality risk management into quality systems. In: Saghee MR, editor. Achieving quality and compliance excellence in pharmaceuticals: a master class GMP guide. New Delhi: Business Horizons; 2012. p. 89–128.

[4]   The Good Automated Manufacturing Practice (GAMP) guide for validation of automated systems, GAMP 4 (ISPE/GAMP Forum, 2001) (http://www.ispe.org/gamp/).

[5]   Chestnut W, Waggener JW. Good laboratory documentation: an introduction. J GXP Compliance 2002;6(3):18–23.

[6]   Agalloco J. Validation: a new perspective. In: Medina C, editor. Compliance handbook for pharmaceuticals, medical devices, and biologics. Boca Raton, FL: CRC Press; 2004. p. 85–128.

[7]   Sandle T. Qualification and validation. In: Saghee MR, editor. Achieving quality and compliance excellence in pharmaceuticals: a master class GMP guide. New Delhi: Business Horizons; 2012. p. 169–206.

[8]   Dolman J. A validated process? GMP Rev 2007;6(2):19–21.

[9]   Chow S. Pharmaceutical validation and process controls in drug development. Drug Inf J 1997;31:1195–2010.

[10]   O'Donnell K, Greene A. A risk management solution designed to facilitate risk-based qualification, validation, and change control activities within GMP and pharmaceutical regulatory compliance environments in the EU. Part 1: fundamental principles, design criteria, outline of process. J GXP Compliance 2006;10(4):12–35.

[11]   Shintani H. Validation study and routine control monitoring of moist heat sterilization procedures. Biocontrol Sci 2012;17(2):57–67.

[12]   Nash RA. The essentials of process validation. In: Lieberman MA, Lachman L, Schwartz JB, editors. Pharmaceutical dosage forms, 2nd ed, vol. 3. New York: Marcel Dekker; 1990. p. 417–53.

[13]   Jenkins KM, Vanderwielen AJ, Armstrong JA, Leonard LM, Murphy GP, Piros NA. Application of total organic carbon analysis to cleaning validation. PDA J Pharm Sci Technol 1996;50(1):6–15.

Microbiological data

20.1 Introduction

Data collected is an important part of microbiology. Each laboratory method generates data in some form or another. Broadly, such data is either qualitative (such as the "pass" or "fail" result from the sterility test) or quantitative (such as a number produced from a bioburden test). This chapter considers some aspects of data capture and data analysis. In total, these are wide ranging areas, and the purpose of the chapter is simply to highlight some of the essential points and provide a base understanding of what microbiological is and how it might be handled. Given that limits setting is doubly important, the chapter explains how microbial test limits might be set, based on a consideration of historical data.

Trending is applicable to most microbiological analyses. Raw monitoring data by itself is of little value, and individual high counts are not often meaningful. Thus, voluminous sets of results need to be appropriately analyzed and presented in order to provide trends and appropriate focus; for example, with the environmental monitoring of cleanrooms.

The reporting and trending of data provides an opportunity for the effectiveness of microbiological control and the appropriateness of the monitoring program to be reviewed and modified. Where good control is demonstrated, there may be opportunities to reduce the level of monitoring, thus reducing cost without compromise to product or patient.

Often microbiological data will contain many zero data points, which can present issues in statistical analysis of the data. It is important to select appropriate analysis tools that do not lead to the masking of significant events or trends.

In whichever way data analysis is undertaken, periodic summary reports should be generated and reviewed by a cross-functional team. In presenting data, a combination of graphical and tabulated formats providing a visual representation with clear supporting summary text is recommended.

20.2 Counting microorganisms

Quantitative data are generated through the counting of microorganisms. One of the important tasks required by pharmaceutical microbiologist is the ability to enumerate microorganisms. Counting is required in order to assess the microbial quality of water, cleanrooms, in-process bioburden samples, of raw materials, and so on. The method used to count microorganisms depends upon the type of information required, the number of microorganisms present, and the physical nature of the sample. An important

Pharmaceutical Microbiology. http://dx.doi.org/10.1016/B978-0-08-100022-9.00020-7

distinction is between total cell count (which counts all cells, whether alive or not) and the viable count (which counts those organisms capable of reproducing).

Total cell counts include direct microscopic examination, the measurement of the turbidity of a suspension (using a nephelometer or spectrophotometer), and the determination of the weight of a dry culture (biomass assessment), adenosine triphosphate (ATP) measurements (typically using the enzyme luciferase which produces light on the hydrolysis of ATP), via fluorescent staining, or electrical impedance. Viable counting techniques include the spread plate, pour plate (through direct plating or an application like the Miles-Mistra technique), spiral plating, and membrane filtration. Such methods were presented in Chapters 1 and 7. To the classic can be added rapid microbiological methods, which often produce more data by being able to address the issue of "unculturability" (see Chapter 17).

20.3   Sampling

The objective of taking a sample is so that the sample taken is representative of the population and by examining or testing the sample, then something meaningful can be inferred about the population. A sample is, therefore, a subset of a population selected by a process. The sample size is the number of items (samples) included. For a water system or an air sample, the sample is a proportion of the total collected at a given time point.

In terms of the numbers of samples taken (the sample design), the sample should be representative. If the sample is not representative or if the sample take is not as intended, then sampling error is said to have occurred (although in practice it is very difficult to know if sampling error has occurred) [1].

In terms of the sample being representative, this means that the sample should be of sufficient volume (such as 200 mL of pharmaceutical grade water) or an appropriate number of samples should be taken in order to produce a representative result (e.g., determining how many samples from a give number of containers of a raw material will give a representative result. There are different statistical tools which can be used for this purpose, the most simple being the square root of the number of containers).

The reason, drawing on the water example above, that the sample size is important since it relates to the distribution of microorganisms. Distribution, as a general principle, is discussed below. In relation to sampling, if water contains 1000 bacteria per liter, this does not mean that each single milliliter will contain one bacterium. However, if a 500-mL sample was taken, then the chance of capturing 50 bacteria is much higher than the chance of capturing one bacterium in a 1-mL sample. In this case, it is more useful to consider the volume required so that a reasonable estimate of the microbial population can be obtained.

For example, suppose we wish to estimate the microbial population in 1 L sample. We could test the entire liter. This would itself be time consuming and expensive, and if the liter was of value, the sample would be rendered worthless.

If we what to be 95% certain of detecting a reliable count, and we know, from experience, the mean contamination level, then the following formula can be applied:

$$V = \frac{\ln(1-P)}{\mu}$$

where V is the volume of sample to test, P is the probability of detecting an organism, and μ is the mean contamination rate.

Suppose, $P=0.95$ (as we are seeking 95% confidence, a commonly applied correction factor to most biological data) and the mean contamination rate is 0.022 cfu/mL. Then, the volume required would be 1498 mL (or, in practice, a 1500-mL sample would be tested).

In terms of where the samples are taken from, the general principle is one of the random sampling. This means that the sample is selected in such a way that every sample of the same size has an equal chance of being selected. Random sampling presupposes that the population is well-mixed before sampling takes place. In drawing upon a microbiological example, random sampling would be applied to the sampling of a raw material from several drums.

Random sampling is not desirable in all cases. For sterility testing, for example, the sampling is biased. The bias is so that the samples relate to time and that the samples taken relate to approximately equal moments during the filling of the product (or, in terms of the numbers of containers filled, the samples are taken at equidistant intervals). Here, something different is being measured than if random sampling was used to select the samples. Furthermore, with environmental monitoring, having fixed samples makes trending more meaningful.

Furthermore, an important aspect of the work of pharmaceutical microbiologists is ensuring that the samples taken or submitted to the laboratory have been done so in an aseptic manner and that the containers and storage conditions of the sample have not been adversely affected [2].

20.4   Microbial distribution

The distribution of microbial counts is an important topic for it has a considerable impact upon the trend charts, general statistic, and on the techniques applied for the calculation of alert and action levels. Microbial counts in the environment rarely resemble normal distribution (where a classical bell-shaped curve or binomial pattern is obtained; here the area under the curve is divided into two symmetrical halves). Normal distribution is a phenomenon found in many aspects of physical and biological science (from measurements like human height). Normal distribution is displayed, at its most simples, as a histogram as shown in Figure 20.1.

In practice, the distribution of microorganisms and microbial counts show either Poisson distribution (such as from a water system where microorganisms are distributed randomly) or a marked "skewness," as with counts from a higher-grade cleanroom. Most microbiological monitoring data displays skewness, where the majority

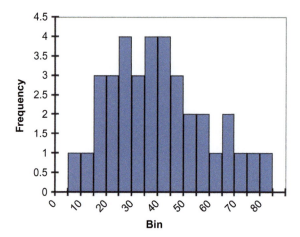

Figure 20.1 A standard histogram, as might be typical for biological data.

of results are zero or low counts, with very few results recording higher counts. Thus, a data plot shows a long, thin tail toward the left of the graph [3].

With Poisson distribution, the frequency of counting "events" over "time" is more random (as with Figure 20.2). Thus, the phenomena of Poisson distribution account for events where a sample may exceed an action level on 1 day, be below it for another 2 days and then be above it again. This situation does not indicate contamination appearing and disappearing, or that one sample has given the correct result and the other has given an unrepresentative one, it merely reflects a distribution across time and space.

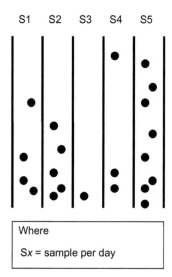

Figure 20.2 Depiction of typical Poisson distribution: a possible distribution of microorganisms in five samples (Sx) taken from the same water outlet.

Outside of statistical parameters, it should be borne in mind that biological data is varied, and there are reasons outside numerical data itself as to why microbiological data is particularly varied [4]. Variations can arise due to several factors, including:

- the monitoring methods (which are often inherently variable);
- culture media, where variations can arise between different types of media (such as general purpose media or fungal specific media), whether the media contains any additives, such as disinfectant neutralizer, and between manufacturers;
- incubation times;
- incubation temperatures;
- sampling procedures;
- sample size or volume;
- different sample locations;
- different sample times;
- frequency of monitoring or sampling;
- the people undertaking the sampling;
- acceptance criteria (such as the means of establishing alert and action levels).

Imprecision in sampling technique can also add to such effects. This fact alone accounts for why a reasonable number of repeat samples should always be taken in response to an out-of-limits event. Such distributions can also enhance the degree of standard error obtained from plate counting.

20.5   Data trending

The reason for requiring the use of trend charts relates to one of the biggest challenges in pharmaceutical microbiology: that is, the vast quantity of data that is collected and the difficulty in interpreting it; and because a single monitoring event only provides a "snap shot" of what may have been happening on a particular day, which may or may not be representative.

Thus, trend analysis is very important for environmental monitoring and is necessary in order for the microbiologist to see "the big picture." There is no right or wrong approach for selecting a trend analysis system, although there are dangers in selecting a trend system that is inappropriate for the data set. Control charts are useful tools for visualizing a process and provide quick "summaries" of process statistics. Control charts make an assumption that the data plotted is independent, that is, a given value is not influenced by its past value and will not affect future values. Control charts have been used for many years as a part of statistical process control (SPC) systems.

20.5.1   Control charts

The lack of normality and seeming randomness of the distribution is of importance when using control charts. Before constructing a control chart, the collected data should be examined to see if it follows normal distribution. Although, as stated earlier, it is improbable that the distribution of microorganisms as indicated by "counts" from environmental monitoring will follow normal distribution, any statistical analysis that is based on normal distribution remains the more accurate approach. Therefore, it is incumbent upon the user to demonstrate first if there is normal distribution, then, if possible, to transform the data to approximate normal distribution. Only when this is not possible should alternate methods be used.

Normal distribution is the basis of the most common statistical methods. This is because [5]:

(a) many naturally occurring populations are normally distributed;
(b) the means of large random samples from populations are commonly normally distributed;
(c) many populations can be made to approximate normal distribution through data transformation (see below).

Normal distribution can be assessed visually using a histogram, blob chart or a normal probability plot (where the resultant plot should lie approximately along a straight line).

However, a problem arises because microbiological data is rarely binomial. Binomial refers to the probability of an event occurring where the event has the same probability of occurring on each occasion, such as, a person being male or female [6]. In contrast, microorganisms in a sample follow Poisson distribution and microbial counts from a test tend to follow a skewed distribution, as discussed above. Hence, microbiologists often need to consider data transformation as a preset to running control charts.

20.5.2 Data transformation

Microbiological data that do not follow normal distribution may need to be transformed for certain statistical techniques and charts. Data can also be transformed to make it easier to visualize them. Furthermore, confidence intervals and hypothesis tests will have better statistical properties if the dependent variable is approximately normal with respect to its mean, with constant variance. Here, transformed data can be used in that normal distribution could not be meaningfully deployed.

Transformation is a mathematic adjustment applied to data in an attempt to make the distribution of the data fit requirements, each data point, which we will call z_i, is replaced with the transformed value:

$$y_i = f(z_i)$$

where f is a function.

In statistics, this is sometimes called "power transform."

Considerations for use for microbiological data:

(a) no one type of transformation is ideal for a particular purposes;
(b) transformations for values of 0 need to be adjusted to a minimum value of 1 (a constant will need to be added to every datum). Remember, when performing other statistical calculations, to adjust the data back by subtracting the constant;
(c) for low count data (where the majority of counts are less than 10). The recommendation is to take the square root to transform the data (alternatively squares can be taken);

(d) for high count data (where the majority of counts are greater than 10). The recommendation is to take a logarithm to the base 10. Logarithmic scales are preferred for large variations in counts. This is because by taking the logarithm of numbers, this reduces the increase in count. Logarithms to the base 10 are used simply because easiest to understand.

In Figure 20.3, some bioburden data are presented. There are three counts of very high values (>1000) against a data set where the mean count is 10.

In Figure 20.4, the same dataset has been used. This time the data has been converted to $\log_{10}$. The data is easier to track.

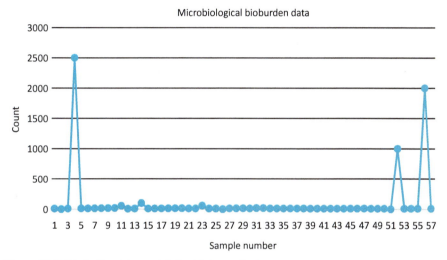

Figure 20.3 Illustrative data, containing high count values.

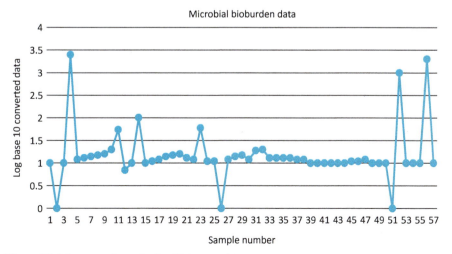

Figure 20.4 Log transformed microbial count data.

Taking the logarithms of numbers also makes the characteristics of data fit better and overcomes the problems associated with non-normal distribution. To assess whether normality has been achieved, a graphical approach is usually more informative than a formal statistical test. For instance, a normal quantile plot is commonly used to assess the fit of a data set to a normal population. Alternatively, rules of thumb based on the sample skewness and kurtosis have also been proposed, such as having skewness in the range of −0.8 to 0.8 and kurtosis in the range of −3.0 to 3.0. A more in-depth explanation is provided by Sandle [7].

20.6   The use of alert and action levels and the setting monitoring limits

Many microbiological methods have specifications, particularly those contained within the pharmacopeia. However, for environmental monitoring, it is typical for the user to set alert and, for nonsterile products, action levels. With environmental monitoring, such levels are not specifications. They are indicators of change and are used for trending purposes and for initiating investigations as required [8].

Alert and action levels can be defined as:

- *Alert level*: a level, when exceeded, indicates that the process may have drifted from its normal operating condition. This does not necessarily warrant corrective action but should be noted by the user.
- *Action level*: a level, when exceeded, indicates that the process has drifted from its normal operating range. This requires a documented investigation and corrective action.

It is important to note that for environmental monitoring, water testing, and other bioburden determinations, we are using the term "level" rather than "limit." Limit is a more absolute term implying pass or fail, accept, or reject. This is appropriate for the sterility test, but meaningless for environmental monitoring where the trend is of prime importance rather than an individual count (which is rarely of significance, due to the variability in microbial counting and the effect of standard error. This is increasingly important when the action level values are small).

There are two approaches for setting alert and action levels. The first is the long established approach of having fixed values where sample results below the value are considered to be satisfactory and sample results at or above the value are considered, as set by both the US Food and Drug Administration (FDA) and within EU good manufacturing practice (GMP), to be excursions. For example, the action level for Grade B/ISO class 7 active air-samples is 10 colony-forming units (cfu)/m^3. Ignoring trend monitoring for the time being, if our results are below the value of 10 they are satisfactory. If our results are at or above 10, they are considered to be unsatisfactory, and an "action" is expected (normally an investigation). For nonsterile facility environmental monitoring, such an action level would need to be set, and for both sterile and nonsterile area environmental monitoring, alert levels need to be calculated.

There are no preset rules for the calculation of alert and action levels. The precise techniques and quantities of data to be used will depend upon several factors, which may include:

- the length of time that the facility has been in use for;
- how often the user intends to use the limits for (i.e., when the user intends to re-assess or re-calculate the limits. Is this yearly? Two yearly? And so on);
- custom and practice in the user's organization (e.g., is there a preferred statistical technique?).

The principles behind the calculation of alert and action levels are:

- they be calculated from an historical analysis of data. The quantity of historical data to be used is something the user will need to define. This can be time based or for a set minimum number of samples. To look at data meaningfully, a reasonably larger number of observations is also required in order for the data set to be representative. So, it is recommended that any analysis is, as a minimum, 1 year or 100 results;
- this should use some type of statistical technique. Statistical techniques are commonly divided into parametric and nonparametric techniques. The difference here is that parametric refers to a procedure that sets out to test a hypothesis about a parameter within a population described by a certain distributional form, which is typically normal distribution. Therefore, parametric methods only really apply to data sets that are normally distributed. An example of a parametric technique is Student's t-test;
- three statistical techniques for assessing monitoring levels are common: percentile cut-off, the normal distribution, and negative exponential distribution approaches.

20.6.1   Percentile cut-off

For low count data (such as EU GMP Grade A/ISO 14644 class 5 environmental monitoring, or from a water-for-injection system), percentile cut-off approach is most suitable (and the easiest to apply). Percentiles are sets of divisions that produce exactly 100 equal parts in a series of continuous values. In order to calculate percentiles, the data must be collected, sorted, and ranked from lowest to highest.

In selecting percentile cut-off values, typically, the warning level is set at the 90th or 95th percentile, and the action level set at the 95th or 99th percentile. Thus, if the 90th percentile is selected, this means that any result above the 90th percentile is 90% higher than values typically collected over the past year (or whatever the data selection period was). There is no easy guide as to the appropriate percentile to select. In this author's experience, it is more common to use the 95th percentile for the alert level and the 99th percentile for the action level. This selection is based on the level of risk that the user wishes to build into the system. A further consideration is whether the microbiologist wishes to round up or down to the nearest zero or five. By doing so, this may make it easier for those using the levels to implement them. Laboratory staff might find it easier to recall and recognize alert and action levels of 10 and 20, as opposed to 8 and 22.

Where data are of a broad range, an alternative approach is to group the data into categories using frequency distribution, for example, 0–10, 11–20, 21–30, and so on. When the category closest to the percentile cut-off is selected, either the mid-point of the category or the upper value of the category can be selected. For example, if the 21–30 category represents the 95th percentile, the action level selected maybe 25 or 30 depending upon the predefined criteria adopted.

To calculate the 95th and 99th percentile, the best way is to copy the data into MS Excel and to use the "PERCENTILE" function. This is:

$$= \text{PERCENTILE}(\text{array}, p)$$

20.6.2   Standard deviations/negative exponential distribution

For higher count data (such as active air-sample counts at Grade D), either standard deviations (if there is normal distribution) or negative exponential distribution (for skewed data) is employed. Unlike the percentile cut-off approach, this technique uses the mean count and observes the spread (or variance) of the different observations. For these approaches, the alert level is equivalent to two standard deviations and the action level to three standard deviations of the mean.

Where the data appears normally distributed, or if a successful data transformation step such as taking the square root or logarithm has been employed, standard deviations can be used to calculate monitoring levels (although there is a danger that inaccuracies can creep in). A common mistake is to produce a histogram and use the second (for the alert level) and third (for the action level) standard deviations. This is incorrect because in doing so this approach has an inherent two-tail probability built into it.

Negative exponential distribution (a term for negatively skewed data) provides reasonable approximation of normal distribution. By multiplying the mean by 4.6, an approximation of the 99th percentile is produced and by multiplying the mean by 3.0, an approximation of the 95th percentile is produced.

20.6.3   Frequency approach

An alternative approach is to use a frequency cut-off approach. This is often applied for aseptic filling operations. The rationale for this approach is that setting alert levels of <1 or 0cfu, with action levels at 1 or 2cfu is scientifically incorrect because neither air-sampling technologies nor current (or anticipated) microbiological methods support these requirements. For instance, there are no standard methods for air sample collection, and variability is comparatively high, based on the metrology and analytical capability of the method. Moreover, there is no data on limit of detection of environmental sampling methods (zero does not mean absence of contamination, it merely means below the level of detection at that point in time) [9].

Another complication is that, at very low recovery levels, there is no agreed way to establish alert or action levels statistically, because the counts are simply too low to make statistical analysis useful. On this basis, it a count of 1 cfu is not significantly different from a count of 10cfu (indeed some scientific literature suggests that ±0.5log is a reasonable assumption of variability).

To overcome this, the second approach to assess areas on the basis of the frequency of distribution of counts and for an "action" to be set if the level of incidences (or "contamination events") exceeds a certain level. With sterile processing areas, the frequency is based on the acceptable level of nonzero counts. Thus, there is an expectation that contamination rate events for aseptic processing should be infrequent.

20.6.4 Problems with limits setting

There are a number of problem areas associated with the calculation of alert and action levels. These include:

- selection of an insufficiently small set of data so that the norm of the process was not captured;
- the set of data was large but for different reasons special causes resulted in it not being typical over a longer period of time;
- whether levels should be lowered or increased following each review if the data set indicates a change in direction. Here it needs to be firmly established if any change in the historical data is due to a special cause or a common cause. If the conclusion is special cause, and these have been corrected, the monitoring levels should probably *not* be changed. Whereas, if changes are due to a common cause the monitoring levels should probably be changed. Even if a common cause is established, it may be prudent not to change unless the trend appears over 2 years. The purpose of reviews should not be to drive limits upwards or downwards without very strong reasons.

It must not be forgotten that by setting monitoring levels based on the premise that 95% and 99% of the data falls within and that 5% and 1% of the data falls without. Therefore, occasional excursions from these levels are to be expected for the data that is gathered and trended over the next year, and some action level excursions will always be expected *if* the data set used for the calculations was truly representative.

It should be noted that for several practitioners, the relevance of having an alert limit is questioned, especially when applied to nonsterile manufacturing. In many instances, if an excursion occurs it will usually go from a "normal level" to action without an increased trend to the alert level.

20.6.5 The need to set monitoring limits

In contrast to sterile manufacturing, here are no major regulatory standards for the setting of limits for nonsterile manufacturing.

The setting of monitoring limits for nonsterile manufacture may involve an assessment of:

- the chemical composition of the product;
- the production process;
- the route of application;
- intended use of the product;
- the delivery system of the product;
- the type of anti-microbial preservative.

When setting limits, many of the principles for sterile manufacturing are relevant.

20.7 Data reporting

Whichever method of reporting is selected, microbiological data must be presented, interpreted, and summarized so that senior management can understand the trend and the "big picture." This data needs to be presented at the correct frequencies (that is not too often or too infrequently). There is no right or wrong way to present data. However, a clear and simple approach is often the most useful. This can include:

- the use of graphs and tables. This allows the trend of one area to be compared with another and for informed questions to be asked;
- each test should be reported separately. Multiline graphs and the use of more than one scale on a graph are generally confusing;
- focus on each filling room or main operation separately. It is often useful to compare different areas, but it is confusing to attempt this on one graph;
- include all of the available data. It is important to select the time period over which the data should be collected and plotted (typically, this will be monthly, quarterly, yearly or, occasionally, over a longer term). Once a time period has been selected, data must never be excluded;
- include warning and action limits on graphs. A trend can sometimes be misleading. It is important to understand how a trend relates to the monitoring levels applied;
- include appropriate information with tables and graphs. This helps to identify patterns and possible reasons for a given trend. Such information includes:
  1. locations,
  2. dates,
  3. times,
  4. identification results,
  5. changes to room design,
  6. operation of new equipment,
  7. shift or personnel changes,
  8. seasons,
  9. heating, ventilation, and air conditioning (HVAC) problems (e.g., an increase in temperature).

20.8 Conclusion

The chapter has considered the variables associated with microbiological data and has outlined how such data is of a skewed distribution and how it does not lend itself to straightforward statistical analysis. Knowing this variation is essential for the interpretation of data relating to microbial counts. This is also necessary for running control charts. There are three general approaches for control charts: histograms, cumulative sum charts, and the Shewhart charts [10].

The chapter has then taken this distribution concept and applied it to data trending and to assigning microbial limits (where no compendial or regulatory limits exist.) Data trending is necessary for microbial count interpretation because individual results from microbiological monitoring are rarely of significance when examining the totality of data gathered. This approach is of importance for exercises such as environmental monitoring or water testing. What is of importance is the direction of the trend that the data is taking over time and the alert levels applied to indicate deviations from the norm.

References

[1] Cochran WG. Sampling techniques. 3rd ed. New York: John Wiley & Sons; 1977.
[2] Cundell AM. Microbial testing in support of aseptic processing. Pharm Technol 2004;58:56–64.
[3] Wilson J. Environmental monitoring: misconceptions and misapplications. PDA J Pharm Sci Technol 2001;55(3):185–90.
[4] Richter S. Product contamination control, a practical approach bioburden testing. J Validation Technol 1999;5:333–6.
[5] Stephens MA. EDF statistics for goodness of fit and some comparisons. J Am Stat Assoc 1974;69:730–7.
[6] Sokal RR, Rohlf FJ. Biometry: the principles and practice of statistics in biological research. New York: W.H. Freeman and Co.; 1995. p. 411–22.
[7] Sandle T. An approach for the reporting of microbiological results from water systems. PDA J Pharm Sci Technol 2004;58(4):231–7.
[8] Tang S. Microbial limits reviewed: the basis for unique Australian regulatory requirements for microbial quality of non-sterile pharmaceuticals. PDA J Pharm Sci Technol 1998;52(3):100–9.
[9] Ackers J, Agallaco J. Environmental monitoring: myths and misapplications. PDA J Pharm Sci Technol 2001;55(3):176–84.
[10] Klein M. Two alternatives to the Shewhart chart. J Qual Technol 2000;32(4):427–31.

Auditing the microbiology laboratory

21

21.1 Introduction

The optimal way to maintain good standards of quality across the microbiological function is through conducting audits. Audits can apply to the laboratory or to associate activities like the activity of sampling. Audits can include those from regulators, accrediting bodies, potential or existing external customers and from the internal quality team. They may be experienced microbiologists, quality system experts or non-microbiologists. Given the wide scope of the subject matter, this chapter is focused on the expectations surrounding an internal audit of the microbiology laboratory. An important aspect of audits, not addressed here, is with the microbiologist going into the process area to assess contamination risks and hygiene practices. The various chapters in this book provide ample material that can be drawn upon by the pharmaceutical microbiologist when investigating production processes and cleanroom activities.

Auditing is a part of Quality Management (which, in turn, is a component of Good Manufacturing Practice). Quality Management is a set of principles, many of which are captured in the ISO 9000 series of quality standards [1]. ISO 9000 captures definitions and terminology relating to quality systems and also links to ISO 9001, which details the actual requirements of a quality system [2]. The ISO standards apply across a range of industries and, whilst useful as an audit tool in terms of structuring an audit, they are not sufficiently process or laboratory specific to be used without additional microbiological knowledge.

There are varying definitions of an audit. Here it is defined as a systematic and independent examination to determine whether quality activities and related results comply with planned arrangements. In addition, the audit assesses whether these arrangements are implemented effectively and are, in fact, suitable to achieve objectives.

According to Martin [3], the key elements of an audit are:

- That it is systematic,
- That it is conducted by someone independent of the operations,
- That the audit be documented,
- That the audit findings be evidenced based,
- That the auditor evaluates the findings,
- That the auditor makes a decision regarding the extent that the audit criteria have been fulfilled.

Undertaking audits is a regulatory expectation, and the process of carrying out periodic audits is in itself an essential part of preparing for regulatory inspections. With this regard, auditing a microbiology laboratory follows the same process as auditing any other

Pharmaceutical Microbiology. http://dx.doi.org/10.1016/B978-0-08-100022-9.00021-9

quality functions. However, to be completely effective, the task requires knowledge of the area being examined—in this case microbiology—on the part of the auditor [4].

This chapter outlines the essential elements for conducting an audit for the microbiology laboratory and features some of the important areas which the auditor should focus on. These areas include:

- Media preparation and quality control,
- Maintenance of microbiological cultures,
- Maintenance of laboratory equipment,
- Laboratory layout and operation,
- Training of personnel,
- Documentation,
- Maintenance of laboratory records,
- Interpretation of assay results.

21.2   Quality audits

A system for self-inspection (alternatively called an "internal audit" to distinguish the audit from a supplier audit) must be in place. Internal audits should be conducted in order to monitor the implementation and compliance with Good Manufacturing Practice (GMP), as well as what it is termed Good Control Laboratory Practice (GCLP) within Europe and Good Laboratory Practice elsewhere. Internal audits should be conducted in an independent and detailed way by designated competent persons.

There are broadly two types of audits that could apply to a laboratory: the compliance audit and the quality audit. A compliance audit is designed to determine whether or not specific activities have been performed according to documented procedures (SOPs). In a compliance audit, the SOPs are not questioned. The objective is to determine compliance with the rules, and the outcome is usually binary, either passed or failed.

However, a good GMP system will not function or improve without adequate audits and reviews. Here the quality audit introduces an element of questioning. A quality audit focuses on identifying the underlying cause of quality problems. Such audits can be very effective in identifying testing practices that may not be as effective as they could be. Whichever class of audit is agreed upon, audits should be performed regularly and systematically, to ensure the laboratory is meeting its objectives [5].

There should be a schedule for carrying out audits, with different activities possibly requiring inspection at different frequencies. An audit should not be conducted with the aim of revealing defects or irregularities. The appropriate philosophy is that the audit sets out to establish facts rather than simply finding faults. Audits often indicate necessary improvement and corrective actions, but they must also determine if processes are effective and that responsibilities have been correctly assigned. The emphasis should be on process improvement and enhancing customer satisfaction.

The generic steps involved in an audit are [6]:

- Initiation,
- Scope,

- Frequency,
- Preparation,
- Review of documentation,
- The programme,
- Working documents,
- Execution,
- Opening meeting,
- Examination and evaluation,
- Collecting evidence,
- Observations,
- Close the meeting with the auditee,
- Report,
- Preparation,
- Content,
- Distribution,
- Completion,
- Report,
- Submission,
- Retention of the audit report.

The above can be simplified as PLAN–DO–CHECK–ACT.

21.3   Auditors and the audit process

Auditors should be knowledgeable of microbiology and be trained specifically as auditors. Auditors should have complete independence of the functions they are auditing. In addition to understanding the processes, strong interpersonal skills are critical to the success of an auditor and the audit he or she is performing. There is a natural defensiveness which occurs on the part of the auditee, and an audit can be an emotional experience. Good auditors are persistent without being relentless; ensure they have their questions answered and thoroughly understand a situation before they evaluate what they have seen and heard.

Some the interpersonal skills required by auditors include:

- Objectivity,
- Tact,
- Fairness,
- Not having any preconceptions,
- Thoroughness,
- Persistence,
- Technical knowledge,
- Strong questioning and interviewing skills,
- Detailed understanding of Good Manufacturing Practices and clauses,
- Confidentiality.

As well as looking for negative points, auditors should highlight positive aspects of the area visited.

In terms of approaching the audit, one generalized approach is:

(a) Initiation

The audit must be initiated by one party, and the two parties (auditor and auditee) must agree to the audit taking place and to the schedule.

(b) Scope

The activities that are to be included in the audit must be decided in advance of the audit by the auditor.

(c) Frequency

The frequency of the audit (how often an area is to be audited) must be determined. In one sense, this will help to determine the scope, for if an area is subject to many audits (such as internal monthly audits) then this may lead to mini-audits focusing on defined parts of the larger area taking place rather than one large audit of an entire facility.

(d) Preparation

Before an audit is undertaken, the auditor must prepare for the audit and understand the function of the area to be audited and its key operations. This includes familiarity with the facility, understanding the type of product produced and how it is organized by personnel and function. This is sometimes referred to as the audit task plan.

(e) Review of documentation

Often the auditor will read documents pertaining to the area to be audited in advance. This may include, depending upon the scope of the audit, the Quality Policy or Site Master File.

(f) The programme

The auditor should map out the programme for the audit in advance. This may include considering questions like: What will be covered? What needs to be looked at? What should be witnessed? Which key documents will be reviewed?

In doing so the auditor will decide upon how the audit will be executed.

(g) Opening meeting

Audits begin with a meeting. The meeting will consist of the auditor and representatives of the area to be audited. At the meeting, the auditor will outline the scope for the audit.

(h) The visit

An "audit" is typically divided into two key steps: visiting the area(s) subject to the audit and documentation review (see below). The visit consists of the auditor being escorted around the area. During this phase, the auditor will examine the area (noting things like fabric), watch activities being carried out, ask questions of staff working in the area, and check records. The auditor may ask for other documents relating to the activities to be reviewed later.

It is a good idea for the auditor to use a checklist. The auditor should take note of things which do not seem correct. These will later be classified into observations and into non-compliances.

(i) Documentation review

Either during the facility tour or at the end of the tour, the auditor will review any requested records. These may include Standard Operating Procedures and training records. Documentation represents an important part of the audit.

Figure 21.1 Simplified audit evidence collection process.

With the visit and documentation review, a useful approach is outlined in Figure 21.1.

(**j**) Close the meeting with the auditee

At the end of the audit, the auditor will prepare the findings. These must be presented to the auditee in a closing or wrap-up meeting. The auditor will convey the importance of the findings and make reference to regulatory documentation where necessary. It is important that anything to be covered in the final report is included in the closing meeting, for the report should not contain non-compliances not raised with the facility staff at the end of the audit.

(**k**) Report

Shortly after the completion of the audit, the auditor must write a report. The report should be held by the auditor's company. Either the full report or, more normal for an external audit, a summary report, is sent to the manager of the area which has been audited. Where non-conformities have been noted, the area which has been audited should respond to the audit findings within a short time period (typically within 30 days). As appropriate, corrective and preventative actions should be set.

21.4   Auditing the microbiology laboratory

The above part of the chapter presented a general model of the audit process. This part of the chapter outlines some of the important aspects of the microbiology laboratory that should be assessed during an audit. The applicability of each topic will depend upon the remit of the laboratory.

21.4.1   Microbiology personnel

During the audit a question may be asked whether the laboratory has the correct number of appropriately trained staff. It is important that written procedures define the training of microbiological staff in both cGMPs and microbiological techniques. This will extend to cleanroom entry and the taking of environmental monitoring samples, in addition to standard laboratory techniques. The laboratory should be able to present a training plan and procedure, together with training records for all staff.

21.4.2   Laboratory design and sample flow

In current cGMP guidelines, it is implicit that microbiology laboratories should be of an appropriate size to carry out all their functions. This needs to cover all aspects of sample handling and storage, dedicated test areas for different materials, media preparation, incubation, handling of microbial cultures and treatment and removal of bio-hazardous waste. Some aspects of work may take place in demarcated areas, such as the reading of environmental monitoring samples, endotoxin testing, and for the identification of microorganisms. Sterility testing must always be conducted within a purpose build test facility.

In addition to the design of the laboratory, auditors will expect to see a logical flow of samples through the testing process and adequate controls to protect the samples and activities from extraneous contamination. For example, the use of qualified unidirectional airflow cabinets and biological safety cabinets as measure to ensure contamination control.

21.4.3   Sample handling

Sample receipt, handling, storage and documentation will be scrutinized. A sample should be labeled with an identity, source of the sample, quantity, batch number or other distinctive code, date sample taken, and date sample received into the laboratory. Additional documentation may be available to define what tests are to be performed on the sample, storage conditions, timescale for test completion or other information. Many microbiological samples have strict storage conditions, and many will have ex-piration dates, and the audit will need to verify these. For example, water samples will need to be placed under refrigerated conditions a short while after sampling and will need to be tested within a defined period (measured in hours) after sampling. Sample handling will either be captured through paper records or via a computerized system.

21.4.4   Culture media

The quality of work undertaken in the microbiology laboratory depends on the quality of the culture media. Therefore, media preparation and quality control form essential elements. As with standard laboratory practice, for media, reagents, chemicals and so forth entering the microbiological laboratory, there needs to be in place procedures that define receipt, preparation, labeling, storage and use.

All containers storing such materials should be labeled with the contents, concentration (if applicable), date of preparation, expiry date, analyst identification and storage conditions. In the case of media, a full documented history or batch record of the process needs to be available. This should include batch of media used, water type, analyst responsible, records of weights, autoclave cycles records, pH adjustments (including pH meter used), growth promotion (if applicable) and a specific batch number. It is expected that the sterilizing process for any media or reagent will be fully validated. If the media has been externally purchased, certification relating to these parameters must be available.

21.4.5   Reference standards

Microbiological reference standards include antibiotic standards, endotoxin standards, biological indicators and microbial cultures. All of these standards need to be controlled in terms of receipt, storage, preparation and use. Such standards are normally accompanied by a certificate of analysis, and such documentation must be reviewed and be made available. As an example, with endotoxin standards, the certificate of analysis defines the potency of the endotoxin, and thus, the volume required for resonstitution.

21.4.6   Control of microbial cultures

Microorganisms are amongst the most common standards used in the microbiology laboratory. The control and maintenance of stock cultures is critical. Cultures should be controlled from receipt through storage and use, to safeguard the purity and identity of the cultures [7].

During an audit, the preparation, handling and storage of cultures will come under scrutiny. This relates to concerns about misidentification and the potential of viability loss through incorrect storage. For long-term storage there should be no more than five transfers (passages) from the original culture. This is necessary to avoid phenotypic variations.

21.4.7   Documentation and electronic systems

The actual control and information for sample handling may follow a paper based or electronic (laboratory information management system—LIMS) system. If it is an electronic system, it should ensure that where appropriate, any requirements for system validation are met [8]. The auditor will focus on password access and the audit trail.

Log books, equipment files and laboratory notebooks will be subject to inspection. A related, and important area of documentation, is training files. These are invariably called upon during an audit and here each member of the microbiology laboratory must be able to demonstrate that they are trained and assessed in the tasks that they regularly undertake.

21.4.8   Laboratory equipment

The quality of microbiological test data is often dependent upon the performance of instrumentation (for example, an incubator or an air-sampler). It is expected that an inventory of all equipment be available, describing instrument type, serial number, date of introduction into the laboratory, calibration and validation processes, and a preventative maintenance scheme.

Within the laboratory it is expected that the status of the instrument is verified, for example by labeling whether it is in maintenance or within calibration. A log book for each piece of equipment should be in place to describe what samples have been tested on or with the equipment, what maintenance and so on has been performed. This is helpful in investigations following instrument or sample failure. Auditors will often request calibration certificates, and it is important that these have been fully reviewed and approved.

An example of an important item of equipment is the laboratory autoclave. Autoclave load patterns should be established, assigned to cycles, and tested. Tests should include heat penetration studies and biological challenge tests.

A common audit finding is for equipment service and calibration reports to state that the equipment was adjusted or was actually out of calibration prior to the service and calibration activity without any follow up being recorded.

Different types of equipment will require different levels of validation and calibration, and the criticality of the instrument will determine this in conjunction with a review of the URS. It is essential that items such as water baths, incubators and refrigerators undergo temperature mapping and that during qualification and routine operation thermometers or temperature probes are traceable to national standards.

21.4.9 Cleaning and decontamination

It is important that the results from the microbiology laboratory are representative of the samples being tested and are not the result of cross-contamination. One way to avoid this is through the cleaning and disinfection of premises and equipment.

In addition, if the laboratory has a media preparation area, this must be well maintained. Media preparation requires the use of mixing containers, glassware, balance weighing boats, spatulas, etc. Written, validated procedures should be available for the cleaning and storage of such components.

21.4.10 Specific tests

An auditor will have specific questions to ask about certain laboratory tests. The purpose of this section of the chapter is not to provide an overview of every test, but instead to draw on some illustrative examples.

21.4.11 Microbial limits testing

Microbial limits testing was addressed in Chapter 7. In relation to audits, an auditor may wish to review:

- Growth promotion and suitability using <100 CFU of each individual organism as an inocula level;
- Requirement for comparison to the previous batch of media;

- Addition of 'nutritive' and 'selective' that is the need to demonstrate that media not only recovers the 'target' organism but also inhibits other organisms;
- Enrichment schemes, for example, for *Escherichia coli* and *Salmonella* the test uses soybean-casein media instead of lactose broth;
- Are the test organisms incubated at the correct temperature?

21.4.12   Preservative efficacy test and antibiotic assays

Where preservatives are present in pharmaceutical preparations, the validation of neutralization is important for routine testing. During product development and stability programmes, there could be a requirement to test the effectiveness of the preservative system.

Antibiotic assays should be conducted according to written procedures. Specific requirements that would be reviewed during an audit include media preparation, temperature control (and hence incubator qualification and monitoring, etc.), control of reference standards and assay procedures, for example, plate design (such as Latin square), concentration and volumes of material used. A thorough understanding of the calculations involved in determining potency would be required. It is insufficient for the analyst to simply put the data into a computer programme and accept the result without knowing the background to the evaluation process.

21.4.13   Sterility testing

In an audit, the auditor would look closely at the sterility test validation process. The facility in which the test is performed will come under scrutiny since it is expected to be of a similar quality to that in which the product was manufactured. Therefore, people and material flows would be as defined for the manufacture of sterile products, and the test would be performed under unidirectional conditions (within cleanrooms) or in isolators. Auditors will assess the techniques for transferring samples, media and test equipment into the test environment. Auditors will also review environmental monitoring results, sterility test failure results and the results of test controls, each of which partly informs about the status of the test environment.

21.4.14   Microbial identification

Procedures should be in place for microbial identification methods. It is important that the laboratory can demonstrate that the staff conducting the identifications have a suitable level of knowledge and experience. An auditor will check that appropriate controls are being run with each identification test session.

21.5   Conclusion

As this book has demonstrated, the scope of works undertaken in a typical microbiology laboratory embraces more areas than those described above. The examples are illustrative and are designed to provide some indication of what an audit may cover.

Quality audits serve an important function within pharmaceuticals and healthcare, not least for ensuring that appropriate standards are maintained. The microbiology laboratory plays an important role within the organization, not least for making decisions about the safety of medicinal products. It is therefore important that the highest standards are maintained, and the audit process can assist in helping to meet this requirement.

References

[1] ISO/FDIS 9001. Quality management systems—requirements. Geneva: International Standards Organisation; 2000.

[2] Tsim YC, Yeung VWS, Leung ETC. An adaptation to ISO 9001:2000 for certified organisations. Manag Audit J 2002;17(5):245–50.

[3] Martin A. Auditing a QC microbiology laboratory. In: Saghee MR, Sandle T, Tidswell EC, editors. Microbiology and sterility assurance in pharmaceuticals and medical devices. New Delhi: Business Horizons; 2011. p. 237–58.

[4] Sharp IR. Quality audit and quality system review in the laboratory. In: Snell JJS, Brown DFJ, Roberts C, editors. Quality assurance: principles and practice in the microbiology laboratory. Colindale: Public Health Laboratory Service; 1999.

[5] Kilshaw D. Quality assurance in clinical laboratories. IMLS Gaz 1991;35:253–5.

[6] Sandle T, Saghee MR. Auditing cleanrooms. In: Saghee MR, Sandle T, editors. Cleanroom management in pharmaceuticals and healthcare. UK: Euromed, Passfield; 2012. p. 551–66.

[7] Manuselis G, Rausch M, Wilson P. Quality assurance in clinical microbiology. Clin Lab Sci 1989;2:34–6.

[8] Sutton S. Qualification of a contract microbiology laboratory. J Validation Technol 2010;16(4):52–9.

Microbiological challenges to the pharmaceuticals and healthcare

22

22.1 Introduction

Microorganisms may be introduced into pharmaceuticals via the materials used in their manufacture and through various environmental sources during processing. Once microorganisms are present, proliferation may occur if conditions are favorable. With sterile products, any contamination presents a risk to the patient. With nonsterile products, an assessment of risk is more complex. Here the absolute numbers and organism types are key factors that need to be taken into account.

To minimize risk, there are key control points that can effectively limit bioburden (the number of microorganisms present in or on a material) and control microbial growth. To explore that the suitable control points are in place and operating correctly, a comprehensive biocontamination control strategy needs to be in place. This will center on limiting the potential for contamination during manufacture.

This chapter takes an overview of pharmaceuticals and healthcare in relation to the different sources from which microorganisms might contaminate a pharmaceutical product. The fate of these contaminant organisms is considered together with the consequences of their survival and growth upon the product and, consequently, for the consumer or patient. In taking this overview, the chapter ties together the main themes of the book and places them within the context of ensuring patient and consumer safety. In doing so, the chapter also re-emphasizes the essential role of the pharmaceutical microbiologist [1].

22.2 Microbial risks to pharmaceuticals

The risk to people from spoiled or otherwise adulterated pharmaceuticals has been appreciated for many centuries. However, it was only during the late nineteenth and the early twentieth centuries that the particular role played by microorganisms in this process was understood. Even here, the steps taken to minimize the risk of contamination were relatively gradual, and these steps have progressed in tandem with increased knowledge. For example, it was not until the 1970s that cleanrooms, originally conceived for the development of nuclear weapons, became commonplace within pharmaceutical facilities in order to provide a clean air barrier to protect the product.

Microbial contamination and spoilage of pharmaceuticals will not only alter the esthetic qualities of a product (color, smell, texture, and so forth); such contamination may also render the product dangerous to the user; or it may also nullify any intended therapeutic value of the product. Here the infection risk presented by pharmaceuticals

Pharmaceutical Microbiology. http://dx.doi.org/10.1016/B978-0-08-100022-9.00022-0

varies according to the route of application of the product, the health status of the user and the nature of the contaminating microorganism (to take some of topics first raised in Chapter 2). Thus, those products which are injected directly into blood vessels or tissues (injections and infusions), and those that are applied directly to the eyes and ears (contact lens solutions, eye drops, etc.) represent a greater infection risk than products that are taken orally or applied to intact healthy skin. The infection risks from injected products and eye products are sufficiently great that all such products must be manufactured in such a way that they are completely free from all types of microorganism (i.e., they are classed as "sterile products"). In contrast, oral and topical products may contain a small number of certain types of microorganisms (excluding specific pathogens) [2].

The assessment of pathogenic microorganisms is important for the risk assessment of nonsterile products, and, in general, monitoring microbial distribution and identifying the predominant isolates are part of good manufacturing practices [3]. Those organisms assessed as dangerous to the patient must be assessed and specified. Although indicator organisms can be used, and it is sensible to include such organisms as a part of method qualification, the list of organisms must be based on what could actually present a risk to the product or process. For such analyses, phenotypic or genotypic microbial identification systems are required [4].

In relation to the patient population, individuals who are immunodeficient, either through clinical disease or through use of immunosuppressive drugs are at a greater risk of infection than young healthy individuals. Damaged skin (burns, cuts, spots, etc.) is more easily infected than healthy intact skin, and special care must, therefore, be taken with products that might be applied to broken or burnt skin.

22.3   Microbial challenges to process environments

The environment where pharmaceuticals and healthcare products are prepared presents a potential vector for contamination (as Chapter 16 demonstrates). With environment, examination of air for microbial content can be demonstrated by exposing dishes containing agar media directly to the air (settle plates) or through drawing in a fixed volume of air through a device that can capture a proportion of the microorganisms in the air-stream (volumetric air-samplers) [5]. Microorganisms falling onto the surface of the agar or deposited upon it produce visible colonies on incubation.

It is not straightforward to generalize about the number of microorganisms present in air, since this varies markedly with location. Air samples taken from within occupied buildings will tend to have a greater bacterial content than those taken outdoors; however, outside fungal cells predominate over bacteria. Typically, total microbial counts of air, taken from occupied rooms, vary between 10 and 10^4 microorganisms per liter. High values are likely to be associated with activities such as milling, hay-making, and so forth. In an uncontrolled laboratory environment, counts typically fall between 100 and 500 cells per liter. Numbers would be appreciably smaller for a cleanroom environment with adequate air filtration. This descending level of contamination tends to run to a standard pattern for areas that are unoccupied (or "at rest," to draw upon cleanroom terminology) [6].

Microbial numbers rise with increasing numbers of people in the room and also with the degree of physical exertion that they are undergoing. The wearing of appropriate protective clothing will do much to contain the shedding of skin scales and bacteria to the air. Further protective measures arise from the use of barrier technology. Examples, including unidirectional airflow devices and isolators, of such measures were considered in Chapter 16.

Examination of soil samples using surface per agar counting procedures generally reveals total viable counts in the order of 10^{12} fungi and 10^{14} bacterial cells per gram. Of these, at least a hundred different bacterial and fungal species will be represented. Air and wind blowing from outside a manufacturing environment will therefore carry with it many bacterial exospores and endospores. These will also impregnate normal outdoor clothing and will be carried on the soles of shoes. This reinforces the need for correct cleanroom entry requirements, basic hand hygiene, and appropriate gowning procedures [7]. Furthermore, in order meet microbial control requirements, all the surfaces within the cleanroom, as well as personnel hands need to be disinfected routinely using a variety of disinfectants [8]. Due to theoretical disinfectant resistance, disinfectants used for the microbial treatment of surfaces are typically rotated and have different modes of action. Normally this means the use of a sporicidal and nonsporicidal agent [9].

Total viable counts for water samples typically vary from between 10 and 10^6 per milliliter. Generally the higher counts are obtained from water that has remained static such as in stagnant pools, drains and holding tanks, whereas the lower counts are associated with free-running water such as in rivers and streams where the majority of organisms are found fixed to large particles in the riverbed. Within the pharmaceutical environment, water is treated to render it low in microbial contamination [10]. This is through processes like reverse osmosis and distillation [11] (as outlined in Chapter 10). The microorganisms isolated from pharmaceutical water systems are overwhelmingly Gram-negative [12]. Risks arise from poorly designed or maintained water systems; or in situations where stagnant water is allowed to remain for long periods on the floor.

The practical setting of the cleanroom, microbial challenges to pharmaceutical products can be airborne or waterborne. Airborne organisms might vary in concentration from 10 to 10^4 per liter, depending upon the grade of the cleanroom, and range in size from 1 to 50 μm diameter [13]. Invariably such organisms are carried on rafts of larger particles, like skin detritus. The surfaces of these organisms will generally be dry and bear net negative electrostatic charges. Size and charge may be used to electrostatically remove or physically filter organisms from air supplies. Water-borne organisms can be present in much higher concentrations than airborne cells (around 10^6/mL). They bear a net negative surface charge and range in size from 1 to 10 μm diameter. Since these cells are killed by desiccation, and then every effort should be made to keep manufacturing environments dry. This is the overriding factor to ensuring good environmental control within the pharmaceutical or healthcare environment.

22.4   Sources of microbial contamination

Microorganisms may ingress into pharmaceutical products along with the raw materials used in their manufacture, together with components used for packaging. Such microorganisms might originate from the manufacturing environment (such as from the industrial plant, air, surfaces, and personnel), or they might enter during storage of the product if the packaging is inadequate or faulty. The final, and arguably most severe, microbial challenge to a pharmaceutical product is administered at the hands of the consumer or healthcare provider. Some of these sources are considered below.

22.4.1   Raw materials

Materials of synthetic or semisynthetic origin are, if stored correctly, generally relatively free of microorganisms. Products of natural origin, extracted from or made of animal tissues (including enzymes, growth factors, gelatine, lanolin, etc.), and plant materials (such as gums, leaf, and grain and root/tuber starches) will carry with them many of the microorganisms normally associated with the living plant or animal. Minerals that are mined and extracted will often be contaminated with organisms from the soil. Since these will include bacterial endospores, and then they will often survive the heat treatments involved in subsequent processing. It is inevitable that some of these contaminants may be potentially hazardous to a product, and they might represent an infection risk to the user. Generally, quality control procedures will determine the numbers and types of organisms that contaminate each and every batch of such raw materials.

A satisfactory batch of a raw material prepared to a pharmaceutical grade should be free from specified pathogenic bacteria, and the total bacterial and fungal population will be below a predetermined value appropriate to the particular product. This is assessed through microbial limits testing, as outlined in Chapter 8. It must be appreciated, all the same, that given suitable conditions bacteria might be able to grow, and multiply in such raw materials. It is, therefore, imperative that the raw materials are well packaged and that suitable storage conditions are maintained and shelf lives strictly adhered to. One of the most significant risks arises from the raw material becoming wet.

22.4.2   Water

Water is a major raw material used not only in the manufacture of pharmaceutical products but also in the cleaning of the manufacturing environment and production machinery. In addition, it is a primary requirement for the growth of microorganisms. Contaminated water probably represents the single-most important source of microorganisms that contaminate products. Gram-negative bacteria, such as *Pseudomonas aeruginosa*, are capable of rapid growth in stored water, even if this is apparently free of available nutrients (such as distilled water). In this respect deionized water often contains high numbers of bacteria which have multiplied within the exchange resins (around 10^4/mL). Even with distilled water (such as water-for-injection), if left for any time at room temperature, then high populations of bacteria can develop (approximately 10^5/mL within 24 h). Water used for the cleaning of floors and plant, or stagnant within the drainage units, "u"-bends, etc., and wet areas within and around taps and sinks are particularly prone to microbial contamination and may act as sources of contaminants.

To avoid some of these contamination scenarios, machinery and plant should be thoroughly cleaned, and excess water removed. Residual water in a mixing vessel, if left, will grossly contaminate the next batch of product. Where possible water used in the production areas should be freshly distilled and presterilized. Stored, piped water should be maintained at a temperature which does not permit microbial growth (circulated and heated), and pipework should not contain dead-legs, traps, or ill-fitting joints where microbes can attach and grow. Moreover, in the production environment, surfaces should be kept dry wherever possible. Since those organisms that grow preferentially in water tend to be Gram-negative, they do not readily survive desiccation.

22.4.3 Manufacturing environment

22.4.3.1 Air

Microbes present within the air of manufacturing environments may ingress into the products where these are open and exposed. Microorganisms will be associated with dust particles drawn in from outside atmospheres, or they may be generated from the shedding of skin-scales by operators and personnel entering the manufacturing facility.

Aerosols created by the turbulent flow of liquids may also give rise to significant numbers of airborne organisms. Thus, the operation of sinks and drains, the running of water taps, and the vigorous cleaning of plant will, at these times, generate airborne organisms.

Ingress of airborne contaminants from the exterior is generally controlled by containing the production environment and providing it with a separate supply of filtered air. Given that the air is sufficiently rapidly recycled within a well-designed cleanroom, and that separate supplies of air protect the filling and mixing equipment, then it is the bacteria introduced along with the clothing and bodies of personnel that require control. Quality standards for controlled environments relate to the air quality (particulate count) of the room and establish working limits on the numbers of microorganisms collected by settle plates and active air samplers.

22.4.3.2 Equipment and facilities

If equipment is inadequately cleaned between uses, or if it is allowed to collect moisture in ports and recesses which directly come into contact with the product, then such processing items can represent potent, and chronic, sources of contamination. Equipment, and pipelines, should be regularly cleaned and stored in a dry state. Appropriate disinfection of exposed surfaces should be employed where appropriate.

If equipment is to be installed or taken into a controlled area, then steps must be taken to decontaminate it or sterilize it before entry. Such procedures would apply equally to maintenance and cleaning equipment used within the area. Withal, walls, and floors within the production environment will accumulate microorganisms and must be subject to regular cleaning and disinfection in conjunction with microbiological monitoring of the facility [14].

22.4.3.3   Personnel

The biggest single source of microorganisms in an enclosed environment arises from the people working within it. Individuals shed many thousands of skin scales to the atmosphere per day. Many of these will be contaminated with skin microorganisms such as yeast-like fungi and bacteria of the genera *Staphylococci* and *Micrococci*. Gowning procedures and hair coverings are intended to minimize such shedding to the environment, and the wearing of facemasks contains aerosols generated by breathing and sneezing.

The axillae of the body and the hair-covered surfaces (armpits, crotch, head, and neck) are ecologically rich niches, and they are especially prone to shedding. Particular attention is, therefore, paid to these areas in the design of specialized clothing for use in the cleanroom and controlled areas. Overshoes prevent the movement of dust borne microorganisms from one area to another. Strict entry procedures must be adhered to, involving hand wash routines, and removal of outside clothing. Personnel must be trained in the use of controlled environments in order to minimize their contribution to airborne contamination. A major objective of such training is to minimize the body movements made by personnel since it is well documented that rapid movements and exertion increase the rate of skin scale shedding and reduce the effectiveness of the protective clothing.

22.4.3.4   Users and healthcare professionals

The final abuse that product has to withstand is that given to it by the consumer or healthcare professional for administration to a patient. If the product is multiuse (i.e., opened, used several times, and stored between uses), then microorganisms may enter it at each use. Such organisms may grow during subsequent storage of the opened product. Furthermore, while storage conditions prior to sale can be adequately controlled those at the point of use cannot be. Products might, therefore, be stored in hot humid environments (bathrooms) where microbial growth will be favored. With multiuse products, the risk is elevated.

Contamination by the user is particularly problematic for creams and ointments that are applied by the fingers and hands. Even with products such as eye drops, it is possible for the user to contaminate the applicator and return organisms to the bottle. In some instances (such as modern mascara dispensers and eye drop bottles), the design of the container is intended to minimize consumer contamination. In other instances, this is not possible. Some products contain preservatives, but the preservative cannot deal with all instances of poor practice.

22.5   Fate of microbial contamination in pharmaceutical products

Once microorganisms have ingressed into a pharmaceutical or cosmetic product, the contaminating microorganisms will either die immediately (especially if the product is then subjected to a terminal sterilization process), survive for some time but be unable to grow and divide (static contamination) or given conditions favorable for growth, multiply (dynamic contamination).

22.5.1   Death of contaminants

If the contaminating microorganisms that ingress into the product die, then the risks to the consumer often die with them. If, however, the microorganisms had produced toxic materials within the product, either before or as a result of death, then these hazards will remain. Good examples of this phenomenon are pyrogens (especially endotoxin, as discussed in Chapter 11).

Other examples of toxic material, not covered elsewhere in this book, are exotoxins and enterotoxins. Exotoxins are secreted by bacteria or released through cell lysis. Exotoxins can trigger toxic shock-like syndrome (TSLS), characterized by hypotension or shock, fever, and multiorgan system involvement. Prominent examples include botulinum toxin (from *Clostridium botulinum*) and texanospasmin (from *Clostridium tetani*) Among the best-characterized exotoxin are superantigens. These are associated with the family of pyrogenic exotoxins produced by *Staphylococcus aureus* and *Streptococcus pyogenes*. The related routes of infection are primarily food poisoning (via ingestion). Although exotoxins have a degree of heat resistance, this is lower than endotoxin. Therefore, in relation to the manufacture of sterile pharmaceuticals, any risks to contaminated glassware intended to be sterile from Gram-positive microorganisms are eliminated through depyrogenation cycles for glass vials [15].

Another bacterial source comes from enterotoxins, such as from *S. aureus*. Enterotoxins are single-chain globular proteins of varying molecular weights. They are released by bacteria and tend to, as the name implies, target mammalian intestines. It is not clear the level of enterotoxins needed to elicit a pyrogenic response, although the levels are again estimated to be considerably higher than endotoxin (at $1 \mu g/kg$) [16]. Monitoring for high levels of staphylococcal bioburden (especially *S. aureus*) would act as a risk control in pharmaceutical manufacturing.

Other cell-related material that can be pyrogenic is peptidoglycan, common to both Gram-negative and Gram-positive bacteria (although it is found in much higher quantities with Gram-positive bacteria). Peptidoglycan is a bag-shaped macromolecule that surrounds the cell. Although peptidoglycan is demonstrably pyrogenic, very high numbers of Gram-positive bacteria are required to trigger a pyrogenic response (at around 10^8 cells; in contrast, with endotoxin, a single *Escherichia coli* cell contains about 2 million lipopolysaccharide molecules per cell and as few as 100 cells could trigger a cytokine reaction in humans) [17].

Death of the contaminating organisms will occur if the products are subjected to a sterilization step such as heating in an autoclave. Death will otherwise occur if the prevailing physicochemical environment is unsuitable for growth (pH, water activity, and so forth), or if there are no appropriate nutrients available.

Changing the solute concentration in the medium also changes the osmotic pressure or the water activity a_w. Thus, water activity is the vapor pressure of a substance divided by the standard state partial vapor pressure of water at a given temperature. This is expressed as

$$a_w = p/po$$

where p and po are the water pressures of the medium and of pure water, respectively, in isothermal and isobaric conditions.

The water activity is described by Raoult's law, so that

$$a_w = f\left[n_w / \left(n_w + n_s \right) \right]$$

where n_w and n_s are the concentrations of water and solute, respectively. It follows that by increasing the solute concentration, the a_w of the environment decreases. As a_w decreases, the thermal resistance of microorganisms increases [18]. This is why freeze-drying is adopted as preservation methods for several types of foods and pharmaceuticals.

If the microorganisms cannot produce dormant-resting states, such as endospores or exospores, then they will die gradually as the intracellular reserves of nutrients are consumed (starvation-induced death). Death in such circumstances can be slow with individual cells surviving in the product for many months after manufacture. The inclusion of chemical antimicrobial agents within the products as preservatives will not only inhibit the growth of most organisms but may also kill some of the static contaminants.

An endospore is any spore that is produced within an organism usually a Firmicute bacterium (here, the classic examples are *Bacillus* species and *Clostridium* species). This is in contrast to exospores, which are produced by growth or budding. Exospore can also refer to the outermost layer of spore in some algae and fungi. Spores can survive under the most extreme conditions for very long periods of time (in theory, indefinitely). Although spores are non-reproductive, once they are removed from environmental stress and nutrient conditions are appropriate, spores are capable of transformation back to vegetative cells, which are capable of reproduction.

22.5.2   Pyrogens and other products of bacterial growth

Gram-negative bacteria release the lipopolysaccharide components of their cell walls into their environment as they grow and divide. Such materials are pyrogenic and are released if the cells are killed. In this instance, release is through degradation of the cells. The amounts of endotoxin released will be in proportion to the numbers of bacteria killed.

The bacterial pyrogens are stable to heating in an autoclave and will if injected into the body of an animal lead to a marked rise in body temperature. Pyrogens must therefore be absent from all injectable products. They can be destroyed by prolonged exposure to high temperatures in a hot-air oven. Terminally sterilized injectable products must, therefore, not contain pyrogens, or high numbers of microorganisms that might liberate them, at the time of autoclaving. Similarly, glassware used for such products must be washed in pyrogen-free water or be depyrogenated, by heating at around 250 °C for 30 min (or greater), prior to use [19]. Other bacterial exoproduct toxins may be allergenic when applied to skin and mucus membranes.

22.5.3   Static contamination

So long as there are viable microorganisms remaining within a product then such organisms have the potential to cause infection in the consumer during product use. If the physicochemical conditions within the product change (such as through the growth of a different organism), then static contaminants may be able to recommence growth. The risks of infection presented by any given organism depend upon:

1. the numbers of bacteria that are likely to be present, and viable, within a single dose (i.e., related to cell number);
2. the route of administration of the product.

Many bacteria are more able to initiate infection more readily and with fewer initial numbers of cells, by the parenteral route than, for example, by ingestion. If sterile products are eliminated from this discussion, then the infection risk is related to the presence of viable opportunistic pathogens in the product. These either will be able to infect by the oral route or will be capable of infecting damaged skin.

22.5.4   Dynamic contamination

A pharmaceutical product is dynamically contaminated when microorganisms present within it are actively growing and multiplying. This will occur when the physicochemical environment is suitable for growth and when adequate supplies of nutrient are present within the product. Some products will be more liable to promote growth than others.

22.5.5   Physicochemical environment

Individual species of bacteria and fungi are capable of growth only within fairly narrow ranges of pH, water activity, and temperature. Within slightly broader ranges of value, growth is inhibited, and the cells survive (so long as sufficient nutrients are available for their subsistence). For obligate pathogens adapted to growth in the human body, the ranges of suitable growth condition are particularly narrow and reflect the homeostatic mechanisms at play in higher animals and plants.

Environmental isolates and opportunistic pathogens are generally able to tolerate much wider ranges of physicochemical conditions. The extreme examples of microorganisms are unlikely to be capable of initiating infection and are equally unlikely to contaminate pharmaceutical and cosmetic products. The most likely dynamic contaminants of products are mesophilic organisms growing naturally in the manufacturing facility and in the water supplies. While endospores and exospores will contaminate products and are generally resistant to heat and chemical agents they are incapable of multiplying as such. Multiplication involves a breaking of dormancy, germination, and successful growth of the vegetative cell before new endospores or exospores can be formed. Thus, spores might be present as static contaminants of a product with extreme pH or water activity but cannot lead to dynamic contamination unless conditions in the product are altered.

22.5.6 Product pH

The pH of a product material is an important factor. Generally, fungi are able to survive and grow at more acid pH (pH 4–6) than the bacteria (pH 5.5–8). Acidic products that are contaminated will, therefore, tend to show signs of fungal rather than bacterial contamination [20]. Since the growth of many bacteria, particularly anaerobic, fermentative organisms lead to the generation of organic acids, and then a product of neutral pH showing dynamic contamination with bacteria will become progressively more acidic.

Eventually acid production will inhibit the growth of the bacteria and facilitate the germination and growth of fungal spores, if present. Dynamic contamination with one type of organisms can therefore lead, at a later stage, to dynamic contamination by previously static organism.

The latter might have been incapable of growth in the unspoiled product. This is referred to as "dynastic" spoilage. In the same manner, the pattern of bacterial growth and spoilage of a product can represent a sequence of bacterial dynasties. Mixtures of microorganisms are commonly found that can spoil a product where individual isolates cannot.

22.5.7 Water activity

Gram-negative bacteria are adapted to low osmolarity environments (high water content) and are, therefore, almost exclusively found growing in water and dilute solutions. Gram-positive bacteria, on the other hand, are by virtue of the strength vested in their cell walls, capable of survival and growth in less dilute systems than Gram-negatives [21]. Most moulds prefer even lower water content environments than bacteria, and some can even generate their own water as a by-product of metabolism. In this manner, dilute pharmaceutical systems (e.g., eye drops, infusions, etc.) will tend to show dynamic contamination with Gram-negative bacteria rather than Gram-positives, yeasts, and fungi. Creams, on the other hand, will, if dynamically contaminated, tend to support the growth of Gram-positive organisms and fungi. Very low water content pharmaceuticals (ointments, tablets) will be relatively safe from dynamic contamination by bacteria but might support the growth of some fungi at the surface of the product if stored inappropriately. This is analogous to the contamination of jams and other preserves where mould will often grow on the surface of the product where water of condensation has formed during filling and where metabolic-water, generated by the fungi themselves, can accumulate.

22.5.8 Nutrients

Microorganisms require certain basic nutrients for growth and maintenance of metabolic functions. The amount and type of nutrients required range widely depending on the microorganism. Microorganisms can derive energy from carbohydrates, alcohols, and amino acids. Most microorganisms will metabolize simple sugars such as glucose. Others can metabolize more complex carbohydrates, such as starch or cellulose, or glycogen found in muscle foods. Some microorganisms can use fats as an energy source.

Amino acids serve as a source of nitrogen and energy and are utilized by most microorganisms. Some microorganisms are able to metabolize peptides and more complex proteins. Other sources of nitrogen include, for example, urea, ammonia, creatinine, and methylamines.

Growth of microorganisms will only occur in a product if appropriate nutrients are provided. In order to grow microbes require sources of nitrogen and carbon.

22.6   Consequences for microbial growth

The previous section considered the possibility of microbial growth. This section discusses what may happen when growth occurs.

22.6.1 Product stability

Microbial growth will lead to reductions in the pH of the formulation. With nonsterile products, this might affect the color of dyes and could cause acid cracking of some emulsions.

With specific product types, excipients such as surfactants can often provide sources of carbon and nitrogen for bacterial growth. If such materials are degraded then they can no longer stabilize the product. Phase separation of oil and water will occur. The texture of creams will be adversely affected by the growth of microorganisms, particularly fungi. The production of gases such as hydrogen sulfide and methane from fermentative metabolism can affect the smell of the product and also cause the creation of gas pockets.

Organic acids produced through fermentative growth of microorganisms can affect both the smell and the taste of products. Many bacteria produce brightly colored metabolites that can drastically alter the physical appearance of the product. Such changes are of particular concern when the product is a cosmetic.

22.6.2 Infection risk

Since the infection risk is related to the numbers of microorganisms, a user is exposed to, then microbial multiplication will increase numbers and hence the degree of risk. This is also associated with the route of administration/use of the product. Spoilage organisms are unlikely to be primary pathogens, but, in high number, environmental contaminants might be able to initiate infection in a compromised patient.

22.6.3 Therapeutic effect

If the source of nutrient to the contaminant organisms are active ingredients of the product, then the therapeutic efficacy of the product will be compromised as the ingredient is metabolized. In this fashion certain antibiotics such as the beta-lactams, some steroid agents and many antimicrobial preservatives (such as parabens and phenolics) can act as nutrients for microbial growth.

22.6.4 Sterile products

All products intended for parenteral, otic, and ophthalmic administration or which might come into contact with damaged or abraded skin (dressings, creams) or mucosa epithelial tissue (bladder) or internal organs (irrigation fluids) use must be manufactured as sterile. Sterility of pharmaceutical products is defined as the total absence in an object or field of all viable forms or life. This is an absolute term, an object or product being either sterile or nonsterile. In conjunction, sterilization is a process that is intended to remove or kill all viable microorganisms from an object or field. Pharmaceutical products may either be subjected to a terminal sterilization process or the component parts may be sterilized separately and assembled into the product and filled into the final containers under aseptic conditions. Aseptic processing is applied to products that are damaged by terminal sterilization processing.

22.7 Microbiological testing

This book has emphasized that the best way to ensure microbial control is through product, process, and environmental controls. It remains that, once the microbiologist has had input into these processes, testing must take place to verify that the product is of acceptable microbial quality.

22.7.1 Nonsterile products

Microbiological tests on the quality of nonsterile pharmaceuticals are limited to estimates of the total viable numbers of bacteria and fungi, together with tests of preservative effectiveness. The latter is performed as a biological challenge test. The fate of known numbers and types of bacteria and fungi inoculated into the product is followed over several weeks. There are no regulatory limits for the numbers of viable organisms that can contaminate a nonsterile product provided that it is free of known pathogens. Most companies, however, set limits of <100 organisms/mL [22].

22.7.2   Sterile products

The microbiological standard set for sterile products is that they should not contain any viable forms of life. This is an absolute term, products are either sterile or non-sterile. "The sterility test" attempts to demonstrate whether or not a particular item is sterile. The basis of the test is that the contents of a product item added to a nutritious medium that will support the growth of any microorganisms within it. Details of the sterility test were presented in Chapter 12. In addition, sterile products must be apyrogenic and free from visible particles.

22.8   Conclusion

Microbiological matters continue to exercise considerable influence on product quality. This chapter pulled together some of the different themes from the book in order to provide an overview of the risks posed from microbial contamination of sterile and nonsterile products. The chapter has also considered some risk mitigation steps, from cleanrooms to antimicrobials; these can either prevent contamination ingress or provide a degree of protection. When assessing such matters, the pharmaceutical microbiologist needs to take a holistic approach.

References

[1] Magdy M. Brief introduction to pharmaceutical microbiologist. J Microbiol Exp 2014;1(5):00029. http://dx.doi.org/10.15406/jmen.2014.

[2] Sandle T. Microbial control of pharmaceuticals. In: Encyclopedia of pharmaceutical science and technology. 4th ed. London: Taylor and Francis; 2013. p. 2122–32.

[3] Akers JE. Environmental monitoring and control: proposed standards, current practices, and future directions. J Pharm Sci Technol 1997;51(1):36–47.

[4] Teske A, Sigalevich P, Cohen Y, Muyzer G. Molecular identification of bacteria from a coculture by denaturing gradient gel electrophoresis of 16S ribosomal DNA fragments as a tool for isolation in pure cultures. Appl Environ Microbiol 1996;62(11):4210–5.

[5] Pasquarella C, Albertini R, Dall'aglio P, Saccani E, Sansebastiano GE, Signorelli C. Air microbial sampling: the state of the art. Ig Sanita Pubbl 2008;64:79–120.

[6] Sandle T. Contamination control: cleanrooms and clean air devices. In: Encyclopedia of pharmaceutical science and technology. 4th ed. London: Taylor and Francis; 2013. p. 634–43.

[7] Sandle T. The use of polymeric flooring to reduce contamination in a cleanroom changing area. Eur J Parenter Pharm Sci 2006;11(3):75–80.

[8] Gen-fu W, Xiao-hua L. Characterization of predominant bacteria isolates from clean rooms in a pharmaceutical production unit. J Zhejiang Univ Sci B 2007;8(9):666–72.

[9]   Chapman JS. Disinfectant resistance mechanisms, cross-resistance and co-resistance. Int Biodeter Biodegr 2003;51(4):271–6.

[10] Sandle T. Avoiding contamination of water systems. Clin Serv J 2013;12(9):33–6.

[11] Underwood E. Ecology of microorganisms as it affects the pharmaceutical industry. In: Denyer SP, Hodges NA, Gorman SP, editors. Hugo and Russell's pharmaceutical microbiology. New York, USA: Wiley-Blackwell; 2008. p. 254–6.

[12] Martino TK, Hernández JM, Beldarraín T, Melo L. Identification of bacteria in water for pharmaceutical use. Rev Latinoam Microbiol 1998;40(3–4):142–50.

[13] Napoli C, Marcotrigiano V, Montagna MT. Air sampling procedures to evaluate microbial contamination: a comparison between active and passive methods in operating theatres. BMC Public Health 2012;12:594–600.

[14] Sandle T. Application of disinfectants and detergents in the pharmaceutical sector. In: Sandle T, editor. The CDC handbook: a guide to cleaning and disinfecting clean rooms. Surrey, UK: Grosvenor House Publishing; 2012. p. 168–97.

[15] Johnson H, Russell J, Pontzer C. Superantigens in human disease. Sci Am 1992;266:92–101.

[16] Brunson KW, Watson DW. Pyrogenic specificity of streptococcal exotoxins, staphylococcal enterotoxin, and gram-negative endotoxin. Infect Immun 1974;10:347.

[17] Pearson FC. Pyrogens: endotoxin, LAL testing, and depyrogenation. New York: Marcel Dekker; 1985, p. 61.

[18] Amaha M. Factors affecting heat-destruction of bacterial spores. In: Proceedings of the 4th international congress on canned foods, Berlin; 1961. p. 13.

[19] Tours N, Sandle T. Comparison of dry-heat depyrogenation using three different types of Gram-negative bacterial endotoxin. Eur J Parenter Pharm Sci 2008;13(1):17–20.

[20] Rousk J, Brookes PC, Bååth E. Contrasting soil pH effects on fungal and bacterial growth suggest functional redundancy in carbon mineralization. Appl Environ Microbiol 2009;75(6):1589–96.

[21] Ryser ET, Elmer MH. Listeria, listeriosis and food safety. 3rd ed. Boca Raton, FL: CRC Press; 2007, p. 173–4.

[22] Beveridge EG. Microbial spoilage of pharmaceutical products. In: Lovelock DW, Gilbert RJ, editors. Microbial aspects of the deterioration of materials. London: Academic Press; 1975. p. 213–36.

Index

Note: Page numbers followed by *f* indicate figures and *t* indicate tables.